KU-788-728

Plate 1 (*see* p. x) Conspicuous, sparkling lighting feature brings festive impression to a public building.

Notes on the Plates appear on pp. x–xx

Lighting design in buildings

Lighting design in buildings

John Boud, B.A., F.Illum.E.S.
Associate, Derek Phillips Associates,
Architects & Consultants
Visiting lecturer, Kingston Polytechnic
School of Architecture

Peter Peregrinus Ltd.

Published by
Peter Peregrinus Ltd.
PO Box 8, Southgate House, Stevenage, Herts. SG1 1HQ, England

First published 1973

ISBN: 0901223 39 5
Library of Congress catalog card number 73–84773

Printed in England by
Billing & Sons Limited, Guildford and London

PREFACE

In the early stages of its preparation, I gave this book the working title 'The design of electric lighting schemes for interior spaces'. The ultimate title is more concise than this, but the original form deserves mention because it does indicate with some precision what the book is about.

It is not a textbook of illuminating engineering or a guide to the selection of lighting equipment on the grounds of technical merit; nor is it a handbook of data and references. It deals with daylighting in its effect on electric-lighting design, but not as an independent topic; the discussion of lighting fittings is intended to help those who select them rather than those who design them. No single volume could usefully claim today to contain 'all you need to know about lighting' and I have assumed that the practising designer has access to the material in Section 1 of the bibliography as well as to manufacturers' literature.

More positively, what I have tried to do is to discuss the essential ideas that must provide the background to the design of lighting schemes. Comprehension rather than comprehensiveness has been the aim. Only the reader, of course, can judge whether the book succeeds in its intention, but this can be put quite explicitly: to leave the architect or interior designer—or a student of these disciplines, or any other interested party—with a better understanding of the principles of lighting design.

These principles, however, must always find particular expression. It may be an exaggeration to say that every problem that arises in practice in designing a building is unique, but it is clearly impossible to have two buildings at the same time on the same site.

Architects and designers may sometimes be guilty of pretentious generalisation over theory, but, when they get down to their real work, it is in a particular context, and their thinking is, I believe, inductive rather than deductive, and rightly so. It is partly for this reason that about one-third of the book is devoted to discussion of lighting in terms of specific building types—a point explored more fully in Chapter 19.

This emphasis on the particular case in lighting as part of building design is related to the difficulty of separating fact and fancy, evidence and evocation. If lighting were no more than applied physics and physiology, design 'according to the book' would produce results that approached the ideal as our knowledge of these subjects advanced. But the subjective and the personal are undeniably important elements in lighting design, and will remain so at least until our understanding of perceptual psychology is very much more profound than it is today. This is added reason for being suspicious of absolute rules and generalised treatments; it also explains why I have not hesitated to write in the first person and without excessive formality. Much of what follows represents a personal view, but I believe that calls for no apology.

Finally, if the book as a whole has a theme, it is that electric lighting has a central importance in the basic design decisions that affect the development of our buildings and our cities. In the phrase that must not be allowed to become an empty cliché, it contributes to the quality of life.

JOHN BOUD

Contents

Part 1

Lighting criteria

Part 2

Lighting equipment

Part 3

Electric lighting in building design

Part 4

Particular building types

TABLES

PLATES

Frontispiece

1 Lighting device built up using Concord's SLZ system for public space in the Sunderland Civic Centre. Photograph by Henk Snoek, London, by courtesy of Concord Lighting International Ltd., London. See also Plate 40

Group 1

between p. 12 and p. 13

2 Lopez Administration Building, Rizal, The Philippines, headquarters of Manila Electric Co. The entrance lobby features a presentation developed by Wheel–Garon Inc., New York, with sculpture by Sara Reid Designs. The main elements, in clear polyester, are 3 m high, and each is lit by nearly 100 light sources programmed to give an hour-long sequence of changing effects. Photograph by Wheel–Garon Inc., New York; print by courtesy of *International Lighting Review*, Amsterdam

3 St. Chad's, Birmingham. The emphasis produced by higher brightness around altar appears as enhanced visual clarity. Photograph by Art–Wood Photography, by courtesy of Philips Electrical Ltd., London

4 Barclays Bank, Floriana, Malta. Architect, Richard England, Malta. Photograph by Gino Theuma, by courtesy of Concord Lighting International Ltd., London. See also Plate 51

5, 6 Showhouse by Wates Ltd., London. Interior and lighting design by Janet Turner. Photographs by Graham-Bishop/Corry-Bevington, London, by courtesy of Concord Lighting International Ltd., London

7 Mullard Works, Blackburn, machine shop. Fittings with specular louvers for precise glare control. Photograph by Hobbs, Offen, London, by courtesy of Philips Electrical Ltd., London

8 Philips Building, Brussels. Landscaped office using recessed airhandling fittings for five 65 W reflector tubes, colour white de luxe. Illumination is 1250 lux, with a uniformity of 74%, from a loading of 69 W/m^2. Glare index, calculated by the British IES system, is 13. Airflow, 80 m^3/h. Photograph by courtesy of N.V. Philips Gloeilampenfabrieken, Eindhoven

Group 2

between p. 28 and p. 29

9, 10 Eleventh Church of Christ, Scientist, London. Original scheme used large indirect fittings for incandescent lamps (1 × 500 W plus 8 × 150 W) on raising and lowering gear. Revised lighting depends on upward light from cold-cathode tubing around the cornice and downlighting from 500 W tungsten fittings serviced from above the ceiling. Lighting design by John Boud. Photographs by Michael Scott, London

11–13 The chapel, Stowe School. Lighting design by Derek Phillips Associates, Bovingdon. Photographs by F. Jewell-Harrison, Bedford, by courtesy of DPA

14 Clement Freud's house at Warbleswick, Suffolk. Semirecessed downlights and track-mounted spots and floods. Photograph by courtesy of Concord Lighting International Ltd., London

15 Stonyhurst School. Lighting design by Derek Phillips Associates, Bovingdon. Photograph by courtesy of the school

Group 3

between p. 44 and p. 45

16 Church at Cuffley. Architect, Clifford Culpin, London; lighting design by Derek Phillips Associates, Bovingdon. Photograph by Henk Snoek, London, by courtesy of DPA

17 Bar in the Hotel President, Russell Square, London. Photograph by John Jochimsen, London, by courtesy of Philips Electrical Ltd., London

18 Lecture theatre in Nijmegen. Design by Dijkema & Croonen, architects. General level of 300–400 lux depends on recessed louvered fittings for three 40 W White De Luxe tubes; 'working wall' and rostrum lit by 30 150 W pressed-glass floodlamps, individually adjustable and controlled in two groups; 750 lux on rostrum, 400 lux on vertical chalk board. Photograph by *International Lighting Review*, Amsterdam

19 House in Hertfordshire. Conversion design by Derek Phillips Associates, Bovingdon. Photograph by F. Jewell-Harrison, Bedford, by courtesy of DPA

20, 21 Fittings by Edison Halo Ltd.; photographs by courtesy of the manufacturer

22 Switching for a board-room installation designed by Derek Phillips Associates, Bovingdon. Photograph by John Maltby, London

23 Commerzbank, Dusseldorf. Bowl-mirrored lamps of 150 W at 950 mm centres in specially designed reflectors; some 60% of the heat is extracted upwards by airflow past the lamp. Design by Hans Dinnebier, 'Licht im Raum', Dusseldorf. Photograph by *International Lighting Review*, Amsterdam

Group 4

between p. 60 and p. 61

24 State Museum of Arts, Copenhagen: Matisse Exhibition. Hall is 11 m high; 150 W PAR38 lamps and 500 W tungsten–halogen floodlights at 3.500 m and 8.250 m above floor level; vertical illumination between 500 and 1000 lux. Photograph by *International Lighting Review*, Amsterdam

25 The library, Trinity College, Dublin. Architect, Ahrends, Burton & Koralek, London, Photograph by Henk Snoek, London, by courtesy of Concord Lighting International Ltd., London

26, 27 Setting in a London exhibition designed by Janet Turner. Photographs by G. I. Smyth, London, by courtesy of the Lighting Industry Federation, London

28 Boys' chapel at Stonyhurst School. Photograph by courtesy of Building Design Partnership, Preston

29, 30 Continuous-trough system developed by Allom Heffer Ltd. Photographs by Dennis Hooker, London, by courtesy of the manufacturer

31 Royal College of Surgeons, London. Architect, Denys Lasdun & Partners, London. Photograph by Henk Snoek, London, by courtesy of Concord Lighting International Ltd., London

32 Main Hall, Cologne Hospital, Bayenthal. Architect, H. P. Tabeling. Photograph by Friedhelm Thomas, by courtesy of *International Lighting Review*, Amsterdam

Group 5

between p. 76 and p. 77

33, 34 Simonbank, Dusseldorf. Architect, Prof. Kraemer, Pfenning, and Sievers, Braunschweig; lighting design by Hans Dinnebier, Solingen. Rows of 40W tubes can be switched in groups; illumination 1100lux. Cylindrical elements have 150mm diameter. Conditioned air enters via slotted diffusing channels and return air is exhausted through openings in the reflectors. Photographs and data by the *International Lighting Review*, Amsterdam

35 The South-Western Electricity Board, Avonbank building is designed on the heat-recovery principle. Air is introduced via linear diffusers and extracted through areas of specular louver between the prismatic panels in the continuous lighting troughs. Average illumination, 1000lux; glare index, 18; tube colour, Natural. Architect and surveyor, Property Section, SWEB. Photograph by Leslie Bryce, London, by courtesy of Thorn Lighting Ltd., London

36 Conditioned air enters via conventional ceiling grilles and is extracted through slots at the ends of the recessed modular lighting fittings. Photographs by Elsam, Mann & Cooper, Manchester, by courtesy of Thorn Lighting Ltd., London

37 Ceiling in landscaped office, Eindhoven. Baffles are 250mm deep and 190mm below a flat 'system' ceiling. Fittings mounted on this ceiling have opal sides and a fine-mesh plastic louver beneath the pairs of 65W White De Luxe tubes. Air is extracted through these louvers. Photograph by N.V. Philips Gloeilampenfabrieken, Eindhoven

38 Ceiling in landscaped office of CIBA Ltd., Monthey, Switzerland. Baffles are 290mm deep and 160mm below the horizontal ceiling. Lighting fittings are recessed troughs with inlet slots

along the edges and extract openings above the tubes. Architect, Suter & Suter, Basle; ventilation, Luwa A.G. Zurich. Photograph by courtesy of *International Lighting Review*, Amsterdam

39 Dining room, Bromsgrove School, Bromsgrove. Photograph by Hobbs, Offen, by courtesy of Philips Electrical Ltd., London

40 Entrance to Council Chamber at the Sunderland Civic Centre. Architect and services engineer, Sir Basil Spence, Bonnington & Collins; electrical installation by Rashleigh Phipps & Co., London. The SLZ light structure supplied by Concord Lighting International Ltd., London; see also frontispiece. Photograph by Henk Snoek, London

Group 6

between p. 92 and p. 93

41 Photograph by Friedhelm Thomas, Campione–Lugano, by courtesy of *Light and Lighting*, London

42 Paramount Cinema, Piccadilly, London. Photograph by Studio Jaanus Ltd., London, by courtesy of Thorn Lighting Ltd., London. See also Plate 63

43 Average illumination, 200 lux; loading, 15·4 W/m^2; tube colour, White De Luxe. Photograph by Christof Hobi, Zurich; print and data by *International Lighting Review*, Amsterdam

44 C & A, Amsterdam; 100 W reflector lamps in recessed fittings, 100 W g.l.s. lamps in metal pendants and 40 W g.l.s. in wall fittings. Photograph and data by *International Lighting Review*, Amsterdam

45 'Gift shop' in de Bijenkorf, Eindhoven, lit by 150 W pressed-glass lamps and 100 W bowl-mirrored lamps. Architect, Gio Ponti, Milan; coarchitect, Th. H. A. Boosten, Maastricht; interior arrangement, Morris Ketchum Jun. & Associates, New York; interior design, Groupe Harold Barnett, Paris. Photograph by *International Lighting Review*, Amsterdam

46 Owen Owen's, Wolverhampton. Photograph by courtesy of Concord Lighting International Ltd., London

47 Woluwé Centre, Brussels. Architect, Marcel Blomme; consulting architect, Copeland, Novak & Israel, New York. Ceiling-mounted fittings are for 150W pressed-glass-reflector lamps at 2.500m spacing. Photograph by courtesy of N.V. Philips Gloeilampenfabrieken, Eindhoven

48 Square louvered fittings, 900mm × 900mm, in suspended ceiling each housing six 20W tubes. Illumination 150–200lux. Photograph by *International Lighting Review*, Amsterdam

Group 7

between p. 108 and p. 109

49 Hemel Hempstead Town Hall: the Council Chamber. Architect, Clifford Culpin & Partners, London; lighting design by Derek Phillips Associates, Bovingdon. Photograph by Henk Snoek, London

50 Cooperative Insurance Society Building, Manchester. Interior by Design Research Unit, London; lighting design by Derek Phillips Associates, Bovingdon; photograph by John Maltby, London. See also Plates 61 and 80

51 Barclays Bank, Floriana, Malta. Architect, Richard England, Malta. Lighting design and equipment by Concord Lighting International Ltd., London, Photograph by courtesy of Concord. See also Plate 4

52 Head office of the Bank of Yamaguchi, at Shimonoseki, Japan. Architect, M. Endo. The relief is lighted by 12 rows of 40W tubes screened by baffles; vertical illumination, 750–1200lux. Design by Prof. K. Matsura of Kyoto University, with S. Matsuda of Matsushita Electric Industrial Co. (lighting research and advisory bureau, Osaka). Photograph by courtesy of *International Lighting Review*, Amsterdam

53 Courts of Justice, Swindon. Photograph by Hobbs, Offen, London, by courtesy of Philips Electrical Ltd., London

54 Interlight House, Feltham, Middx., headquarters of Merchant Adventures Ltd., lighting-fittings manufacturer. Photograph by Colin Westwood, London

Group 8

between p. 116 and p. 117

55 Scottish Stock Exchange, Glasgow: modernised accommodation in a Victorian building. Illumination 400–450lux. Photograph by Ian Hamilton, Glasgow, by courtesy of Thorn Lighting Ltd., London

56 Offices of Philips Lighting Division, Croydon. Photograph by Purchasepoint Group, London

57 Ove Arup & Partners, consulting engineers, London. Photograph by Studio Jaanus Ltd., London, by courtesy of Thorn Lighting Ltd., London

58 National Westminster Bank, Colmore Row, Birmingham. Heat-reclaim system designed by Thorn–Benham Environmental Unit, London. General light in public area with coffered ceiling finished in gold leaf is provided mainly by 250W fluorescent-bulb mercury lamps in semirecessed airhandling fittings. Photograph by F. R. Logan Ltd., Birmingham

59 Ford Foundation headquarters, New Delhi. Architect, Joseph Allen Stein & Associates. Each louvered fitting houses six 20W cool-daylight tubes. Illumination on desks, 600lux. Photograph by courtesy of *International Lighting Review*, Amsterdam

60 Photograph by courtesy of *Light and Lighting*, London

61 Cooperative Insurance Society Building, Manchester. Interior by Design Research Unit, London, lighting design by Derek Phillips Associates, Bovingdon. Photograph by John Maltby Ltd., London. See also Plates 50 and 80

62 South-Western Electricity Board, Avonbank, Bristol. Photograph by Leslie Bryce, London, by courtesy of SWEB

Group 9

between p. 124 and p. 125

63 Paramount cinema, Piccadilly, London. Photograph by Studio Jaanus Ltd., London, by courtesy of Thorn Lighting Ltd., London. See also Plate 42

64 Thorndike Theatre, Leatherhead. Photography by Richard Einzig of Brecht–Einzig Ltd., London

65, 66 Lecture theatre, Oersted Institute, University of Copenhagen. Architect, Mogens Voltelen. Fluorescent and deep-baffled 40 W fittings by Nordisk Solar Comp., Copenhagen; 100 W fittings by A/S Lyfa, Copenhagen. Photographs by Struwing, Copenhagen, by courtesy of *International Lighting Review*, Amsterdam

67, 68 The West Herts. & Watford Postgraduate Medical Centre. Architect and lighting designer, Derek Phillips Associates, Bovingdon. Photographs by F. Jewell Harrison, Bedford. See also Plate 78

69 Photograph by Friedhelm Thomas, Campione-Lugano, by courtesy of *Light and Lighting*, London

Group 10

between p. 132 and p. 133

70, 71 Cox & Wyman Ltd., Fakenham, Norfolk. Scheme uses over 120 1800 mm × 600 mm recessed modular airhandling fittings with opal diffusers. Illumination is 600–700 lux; each fitting houses 4 × 85 W Daylight tubes, and handles 130 m^3/h. Architect, Alec T. Wright, Adlam & Brown, Norwich. Photographs by Colin Westwood, by courtesy of Crompton Parkinson Ltd., London

72 Works of Shin–Nihon Seitetsu Co. at Kimitsu, Japan. Upward light improves environment. Fittings for 700 W mercury lamps mounted at 19·5 m, spacing 15 m. Average illumination (for rough work) 150 lux. Photograph by courtesy of Matsushita Electric Industrial Co, Osaka; print lent by *International Lighting Review*, Amsterdam

73 Herbert Ingersoll Ltd., Daventry. This plant for the design and construction of advanced manufacturing systems is centred on a single-storey windowless building covering about 95 000 m^2; it is lit by 450 1000 W MBF/U lamps in mountings that incorporate ventilation. Ceiling height 10·5 m; illumination 1000 lux. Consulting and electrical engineer, W. S. Atkins & Partners, Epsom

74 H. P. Gelderman & Zn., Oldenzaal, Holland. Fittings for 65 W tubes in rows 1·8 m apart. Mounting height, 4·5 m. Average illumination 400 lux. Scheme by Philips Nederland NV, Eindhoven. Photograph by courtesy of *International Lighting Review*, Amsterdam

75 Machine shop at the Telegraph Condenser Co., Acton, London. Photograph by Hobbs, Offen, by courtesy of Philips Electrical Ltd., London

76 Fittings by Thorn Lighting Ltd. Photograph by Studio Jaanus Ltd., London

77 Photograph by Studio Jaanus Ltd., London, by courtesy of Thorn Lighting Ltd., London. Operating theatre fitting by Sierex Ltd.

78 Architect and lighting designer, Derek Phillips Associates, Bovingdon. Photograph by F. Jewell Harrison, Bedford

Group 11

between p. 140 and p. 141

79 Restaurant on the *Queen Elizabeth* 2. Photograph by courtesy of N.V. Philips Gloeilampenfabrieken, Eindhoven

80 Restaurant in CIS Building, Manchester (see also Plates 50 and 61). Interior by Design Research Unit, London; lighting design by Derek Phillips Associates, Bovingdon. Photograph by John Maltby, London

81 Architect, Prof. Roland Rainer. General light from 108 downlights for 1500 W incandescent lamps, with additional and effect lighting using some 85 parabolic reflectors and about 120 floodlights, including theatre types. Photograph by Elin–Union, Vienna, by courtesy of Pressestelle der Stadt Wien. Print and data from ILR

82 Luminous ceiling of 60 mm square louvers screening more than 2000 65 W De Luxe Warm White tubes. Directional light from 300 W PAR56 reflector lamps. Illumination, 1000 lux. Photograph by H. J. C. A. van Stekelenburg for *International Lighting Review*, Amsterdam

83 Gymnasium for Sunderland Football Club uses 80 400 W 'Kolorarc' mercury lamps; maximum illumination 500 lux. Photograph by Turners (Photography) Ltd., Newcastle, by courtesy of Thorn Lighting Ltd., London

84 Photograph by G. I. Smyth, London, by courtesy of Thorn Lighting Ltd., London

85 Architect, Robert Matthew, Johnson-Marshall & Partners; consulting engineer, Steensen, Varming, Mulcahy & Partners. Photograph by Ian Hamilton, Glasgow, by courtesy of Thorn Lighting Ltd., London

86 The Kyoto conference centre has six major rooms. The ceiling illustrated is of a hall for 550 people. Architect, Sachio Ohtani of Tokyo University. Incandescent downlights take 200 W and 60 W lamps, and the pattern of aluminium beams screens 40 W tubes. Illumination, 200 lux. Photograph by Y. Hayashi, by courtesy of the Matsushita Electric Industrial Co., Lighting Research and Advisory Bureau, Kadoma, Osaka, Japan. Print and data provided by *International Lighting Review*, Amsterdam

Group 12

between p. 156 and p. 157

87 Church at Stonyhurst College, Stonyhurst. Lighting design by Derek Phillips Associates, Bovingdon. Photograph by courtesy of the college

88 Track-mounted fittings are for 300 W PAR56 lamps. Photograph by courtesy of Concord Lighting International, London

89 Commonwealth Institute, London. Baffles screen rows of Trucolor tubes. Photograph by Art-Wood Photography, London, by courtesy of Philips Electrical Ltd., London

90 Skyline Hotel, Bath Road, London Airport (Heathrow). Architect, the Ronald Fielding Partnership. Lighting fittings by Edison Halo Ltd., London, by whose courtesy the photograph is reproduced

91 Photograph by Stewart Bale Ltd., by courtesy of Conelight Ltd., manufacturer of the adjustable fittings shown

Part 1

LIGHTING CRITERIA

Chapter 1

Introduction

When primitive man first made his home in a cave, his motives were probably mixed. Protection was certainly among the advantages this dwelling offered—from wild animals and unsympathetic neighbours, but also from the wind, the rain, and the sun.

In the fullness of time, he discovered that an artificial cave—a building—was, on balance, better. It kept him warm and dry, while a relatively small hole in the roof would let out the smoke if he lit a fire. And, within much wider limits, he could choose where he wanted to live.

After a bit, the *avant-garde* started suggesting that a building might somehow be pleasant to look at, in itself, and not simply through its associations or its decorations (they had all been keen on murals since the cave days). This was a wild and not very relevant idea for most of the tribe, but they did not object to a few intellectuals talking about buildings in this way so long as it did not interfere with their effectiveness.

Intellectuals are often articulate, and, by the time the tribe produced its first colour supplement, it had architectural critics who wrote about buildings first and foremost as though they were some form of monumental sculpture.

It is therefore necessary, at the very beginning of this consideration of lighting in buildings, to state the assumption on which it is based: a building exists to modify the physical environment to make it more suitable for the activity that is being housed.

Take office work, in our society one of the commonest kinds of human activity. What are the objections to doing it in the open air? Snow, rain, mist or fog may not only make conditions positively unpleasant for the people engaged in the work; they will also have an effect on the materials and the machines (paper, typewriters and so on) that will make the work virtually impossible. Conditions will frequently be too cold or too hot for comfort and concentration. Even when there is a pleasantly warm, dry day, the wind tends to blow papers about or to stir up dust. If the office work is going on in its usual location, the commercial area of a town, traffic noise represents a worrying distraction; it may also interfere with conversation essential to the work. Towards the beginning and end of the day in winter and at other times, it may be too dark for work to be possible or to be carried on without strain. On the other hand, the summer sun may make it too bright for the necessary visual comfort. Other disadvantages of the open air as a setting for office work include the dull, enervating effect of an overcast sky, and the possible presence of noxious smells, insects, bacteria and the like.

The point need not be laboured. The spontaneous physical conditions which constitute the natural environment are, to say the least, unlikely to be suitable for a major part of the human activities that make up civilised life. They will frequently be positively hostile, and almost never ideal.

So, buildings exist as the means by which man changes the unsuitable into the acceptable, the optimum, or the ideal for a particular activity, depending on what his technology makes possible and on how much he is prepared to spend.

Looking back at the list of objections to doing office work in the open air, and putting the same ideas positively and succinctly, we may say that the purpose of a building is the production of spaces suitable for the activity where there is some degree of control over environmental factors, which include humidity, temperature, air movement, air pollution, noise and lighting.

One point to be cleared up straightaway is that, when we say 'lighting', we tend more and more to mean not the hardware that produces the effect but the total result; it is a pity that the expression 'visual environment' is long and seems pretentious, for this is really what we are concerned with.

In thinking how this visual environment can be created or controlled, we must pursue four groups of topics:

(a) the criteria that express desirable qualities
(b) the hardware that is available to produce the effects we are seeking
(c) the way that the design of the lighting interacts with other aspects of the total building design
(d) the application of these ideas to particular types of building.

These four 'headings' represent the four parts of this book. In the first, we attempt not so much to evolve a rigid guide or check list, but rather to consider some of the qualities of the visual scene that must be borne in mind when the objectives of any piece of lighting design are formulated. As a tentative generalisation, let us say that, in most interior lighting, the aim is to produce an adequate illumination on relevant planes while limiting discomfort from glare, direct or indirect, to an acceptable extent and achieving a proper distribution of brightness generally; the light sources should have suitable colour rendering, and the scheme should result in appropriate modelling; the degree of uniformity or diversity and the extent of simultaneous or sequential variety should be consistent with the situation; and the whole should be achieved at an acceptable cost, in a way that permits effective and economic maintenance. Such a statement, however, is so riddled with terms that seem to beg more questions than they answer—'adequate', 'relevant', 'acceptable', 'proper', 'suitable', 'appropriate'—that the need to look at the criteria one by one emerges strongly.

Chapter 2

Illumination

We should all, of course, agree with Lewis Mumford that any change of emphasis from quantity to quality represents an advance in civilisation. But the spontaneous first question about most lighting schemes is 'How much light do we want?' The answer depends on a number of factors, the most important being acuity and amenity—terms that appear frequently in the jargon of lighting design.

Visual acuity, or sharpness of vision, depends on how much light there is. We recognise this almost without a second thought; we acknowledge this every time we take something with small detail or subtle contrast to the window or the desk lamp to examine it. Visual acuity can be defined formally, and quantified in terms of the angle subtended at the eye by the smallest detail that can be perceived; its relationship to illumination—the concentration of incident light—can then be studied experimentally; work of this kind has been carried out over the years in many countries. The results show that acuity increases rapidly with illumination at relatively low levels, but that the well known 'law of diminishing returns' applies: in Fig. 1.1, increasing the illumination from A

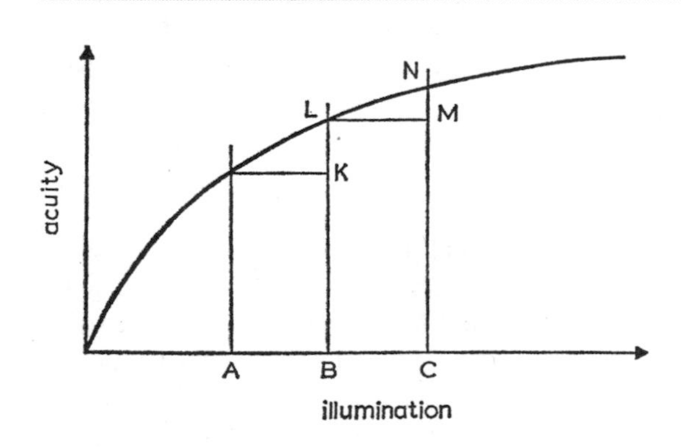

Fig. 1.1 Form of relationship between visual acuity and illumination

to B produces a change in acuity represented by KL, but an equal further increase from B to C leads to a smaller improvement MN.

The units of visual acuity need not concern the practising designer. Illumination is measured in terms of the flow of light arriving on a square metre. If 1000 lumens fall on $5\,m^2$, the average illumination, or illuminance, is said to be $1000 \div 5$, that is 200 'lux' (the approved symbols are 'lm' and 'lx', though there seems little need for the latter).

Some typical levels may be listed:

strong, direct sunlight	50000 lux
dull, overcast sky	5000 lux
drawing office	500 lux
domestic living room	50 lux
good streetlighting	5 lux
moonlight	0·5 lux

The immediately striking thing about these figures is the enormous range—some sort of vision is possible over brightnesses varying by a million to one. We may also notice, however, that the subjective impression of change of brightness is broadly similar in going from 50 to 500 lux (average living room to drawing office) and from 5000 to 50000 lux dull day to strong sunlight). What is important is not the arithmetical difference but the ratio of the levels—in each case here, 10 to 1.

An interesting pattern emerges if we take the common logarithms of the numbers in the list above. The logarithm of 5 is 0·699; log 50 = 1·699; log 500 = 2·699; and so on. In other words, broadly comparable changes in the subjective impression are caused by illumination levels with logarithms changing by a constant amount, in this case 1. Most human sensations depend on a physiological response to a physical stimulus; the relationship of the one to the other is not simple, but, in a number of cases, it is very roughly true to say that the sensation is proportional to the logarithm of the stimulus. In acoustics, changes in sound levels are usually measured and quoted on a logarithmic scale (bels and decibels), and it has been suggested that a comparable notation might be useful in lighting. The steps in the list of typical levels are all of 1 bel, or 10 decibels, and the range from moonlight to strong sunshine is 50 decibels, these numbers agreeing more closely with our impressions than a comparison of 50000 and 0·5 lux. While it seems unlikely that an illumination scale of this kind will be generally adopted, at least in the immediate future, two related points of importance emerge.

The first is that changes of less than 1 decibel do not mean much in human terms. This corresponds to a change in lux of 25–30%; so there is little point in arguing about the difference between 440 and 460 lux. Indeed, it is often said that a 50% increase is the smallest to any real significance, and this is reflected in codified illumination recommendations (200, 300, 450, .. lux) and the wattages of general lighting service lamps (40, 60, 100, 150, .. W).

The second is that the graphs such as Fig. 1.1 are often more usefully drawn with illumination* on a logarithmic scale, i.e. one where equal ratios of lux are indicated by equal lengths along the horizontal axis. If the approximate 'law' mentioned earlier were true, the graph of sensation against illumination level would be a straightline, as in Fig. 1.2. Perceptual psychologists will be interested in how close this may be said to be to truth, and, to pursue it, they have to tackle the vexed question of how to measure sensation. Fortunately for the

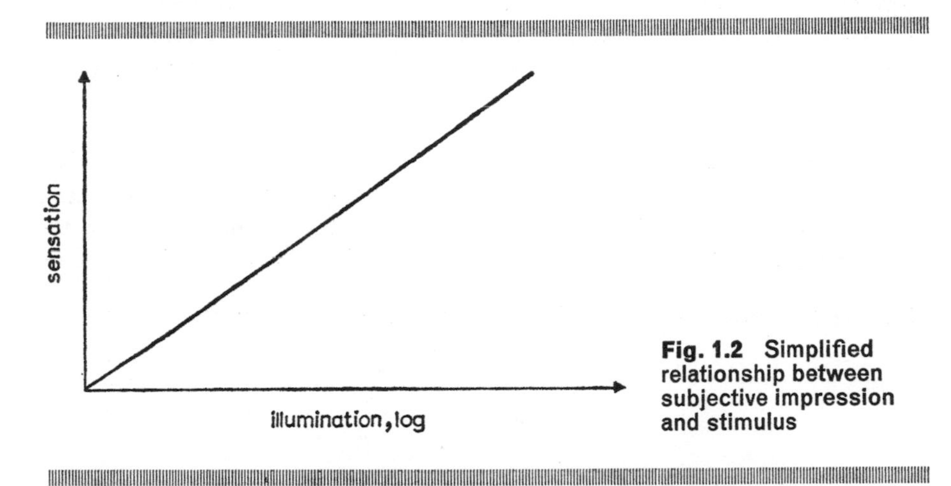

Fig. 1.2 Simplified relationship between subjective impression and stimulus

practical lighting designer, however, his concern is with things that are related to sensation but can be investigated and quantified more readily—visual acuity, the speed and accuracy of carrying out a specific visual task, expressions of satisfaction, and so on. But, for the reasons outlined here, these results are

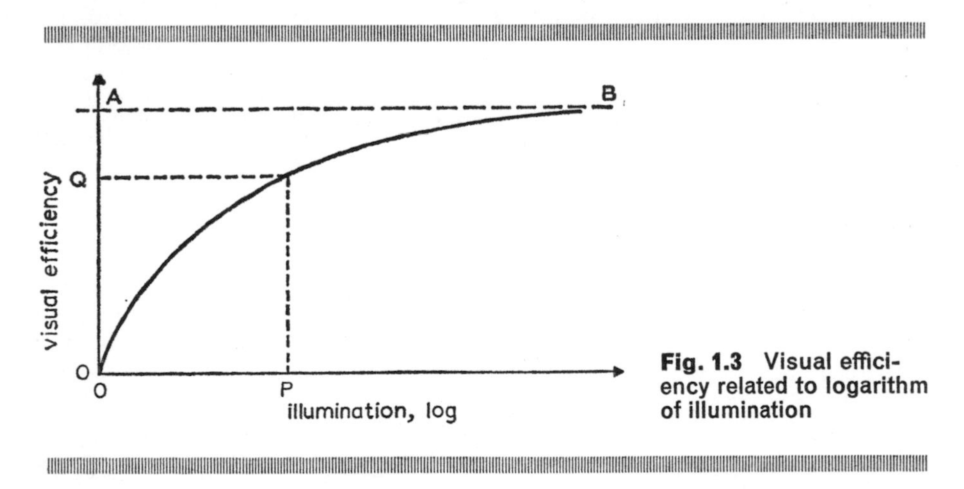

Fig. 1.3 Visual efficiency related to logarithm of illumination

* A term still in use for 'illuminance'; *see* Postscript.

often best expressed as a graph of the quantity against the logarithm of the illumination.

Suppose we plot in this way how 'visual efficiency' changes (whatever the terms may mean precisely). The graph that emerges is still a curve, as in Fig. 1.3. The curve seems to approach ever closer the broken line AB without quite reaching it. We can regard this line as showing the greatest possible visual efficiency (namely the value that cannot be increased by any further change in illumination alone). The efficiency produced by any particular illumination, such as that corresponding to OP, can be read off (OQ) and expressed as a percentage of OA (here, say, 70%).

A similarly shaped curve appears if we are able to plot, in commerce or industry, the monetary expression of the value of the lighting against illumination. Here an ordinary arithmetic scale for lux is more useful, since the cost of providing the lighting will—other things being equal—depend directly on the level provided, and so will be represented by a straight line.

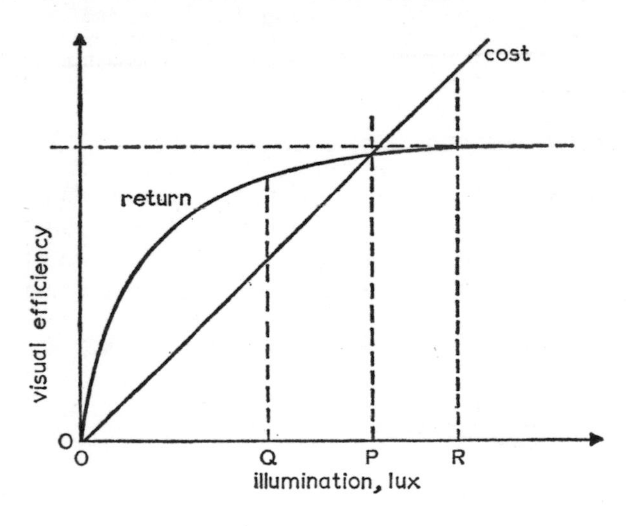

Fig. 1.4 Investment and return related to illumination level

The return from the investment in lighting exceeds the cost when the illumination is represented by, say, OQ. An illumination of OR may be criticised, as the cost of providing it exceeds the return it brings. It is sometimes urged that OP represents the optimum illumination, but the validity of this conclusion can be challenged. However, this does not in the end matter very much, for it is virtually impossible to draw this graph for a particular situation, as many of the quantities involved cannot be worked out with any certainty. The value of sketching the general case, as we have done here, lies in the fact that it shows clearly that an optimum illumination is an economic compromise, an attempt to find the best value for money.

Most national lighting codes rest on an estimate of the percentage of the maximum visual efficiency that, in typical circumstances, represents this economic balance. Fig. 1.5 shows the relationship between visual efficiency and illumination for two tasks of contrasted difficulty. The curve for the easy visual task (large detail and high contrast) rises rapidly, and a particular percentage of

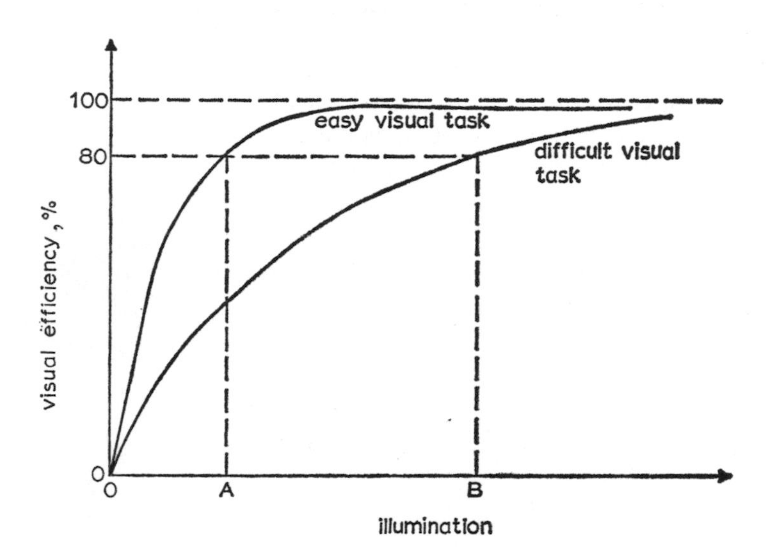

Fig. 1.5 Effect of illumination on easy and difficult visual tasks

the maximum visual efficiency, say 80%, is reached at a relatively low illumination (OA). With small detail and low contrast, a much higher level (OB) is needed to reach the same standard of visual performance. (The graphs in this chapter and elsewhere are intended merely to illustrate principles, and should not be regarded as representing specific numerical results.)

Whereas earlier codes explicitly stated the basis of their numerical illumination recommendations, later documents recognise the fact that the figures they quote express something beyond simple academic analysis. The results of research are considered in relation to the wealth of experience that has produced good current practice. The British IES code of 1968 (Reference 1.3) says: 'In scientific investigations to establish the nature of the relations between these criteria (performance, efficiency and comfort of the worker) and illumination levels, it is found that these relations are necessarily statistical; the point on this statistical distribution to be adopted as the general norm is decided by economic considerations. Recommended illumination values based on these considerations are given for a large number of tasks. . . .'

It is easy to imagine work that makes very modest demands on visual acuity—such as sorting large white boxes from small black ones. Visual efficiency for this task might be 99% at 25lux, but no management today would decide that 25lux was adequate for the interior. For the illumination represents one of the amenities of the environment, and a level of perhaps 200lux would be provided almost solely on the ground that this is where human beings spend their working lives. The distinction between amenity and prestige is blurred, and there exists a whole range of considerations, outside the functional effect on vision, that can influence the choice of illumination level.

This brief discussion of illumination has merely introduced some of the relevant ideas. The subject will recur throughout the book, both in its general chapters and in the sections on particular building types.

Chapter 3

Glare

We reached the tentative conclusion in Chapter 1 that, 'in most interior lighting, the aim is to produce an adequate illumination on relevant planes while limiting discomfort from glare, direct or indirect, to an acceptable extent..' Although most people would probably feel that they understood what it meant in broad terms, we need to look at it rather more closely.

Glare is difficult to define, partly because the word is used for a wide range of effects. Glare may not be recognised as such in a situation where an imbalance in brightness has a cumulative effect; on the other hand, it may be immediately and obviously uncomfortable, or even disabling, in the sense of making seeing impossible.

Conventionally, we distinguish between 'disability glare' and 'discomfort glare'. The former might be caused by undipped opposing headlamps or some other extreme contrasts, but we should not normally expect to meet it in interior lighting. We are concerned with the discomfort caused by brightness and the ways of confining it within acceptable limits—which means, of course, that we must understand what those limits are.

A number of techniques for a systematic check on the glare characteristics of a lighting scheme have been devised in various countries. The British system is outlined here as an example.

Work at the UK Building Research Station (BRS) has established that the glare effect of an area of brightness depends mainly on four factors:

(*a*) the objective brightness (or 'luminance') of the area
(*b*) the apparent size of the area at the viewpoint in question
(*c*) the luminance of the background
(*d*) the position of the glare source in relation to the direction of view.

The published results (discussed in, for instance, References 2.10 and 2.15) propose a formula connecting the glare effect with the numerical expression of these factors. On this formula rests the 'glare index' system developed by the Illuminating Engineering Society (IES), London. Tables have been produced by computer giving values of the glare index for a large number of typical situations, and these form the basis for the routine computation of the index, which is found by applying a series of corrections to the tabulated data. The calculation is described with worked examples in, for instance, 'Interior lighting design' (Reference 1.2). Alternatively, slide-rule calculators are available.

The orthodox view of routine lighting design is that the starting-point is a specified level of illumination in service. An array of lamps and fittings that will produce this level is worked out by the lumen method. It remains tentative until the glare index has been calculated; if it does not exceed the 'limiting glare index' quoted in the IES Code, the scheme is acceptable, but, if it is outside the limit, the proposals must be modified appropriately.

Examples of this design method are given later (in the discussions of office and factory lighting, Chapter 21 and 23). For the moment, it is more valuable to consider, briefly, how glare is related to each of the four factors given above. The glare index is likely to be evaluated only for a restricted proportion of lighting schemes (such as where the brief specifically calls for the IES Code glare-index recommendations to be observed). But an understanding of the causes of glare is valuable in almost every piece of lighting design.

The glare caused by an area of brightness increases very rapidly with that brightness (or, strictly objectively, its luminance). The nonmathematical person may be daunted to see the relationship expressed as

$$G \propto B_s^{1 \cdot 6}$$

but all that this means is that the effect is proportional to the brightness of the source raised to the power 1·6. Assume that, the other factors remaining constant, the luminance of a glare source changes from $1000 \mathrm{cd/m^2}$ to $2000 \mathrm{cd/m^2}$. The glare initially is proportional to $1000^{1 \cdot 6}$, or about 63000, and then to $2000^{1 \cdot 6}$, or roughly 191000; in other words, doubling the brightness has more than trebled the glare effect.

The influence of the other three factors can be expressed briefly. Glare varies with the apparent size of the source (strictly, with the solid angle ω it subtends at the viewpoint) but doubling the size does not double the result; it falls as the background luminance rises (doubling the background brightness halves the effect); and it also falls, quite rapidly, as the angle increases between the source and the direction of view. One expression of the BRS result is

$$\mathrm{G} = \frac{B_s^{1 \cdot 6} \omega^{0 \cdot 8} p}{B_b}$$

B_s is the brightness of the source and B_b that of the background; ω is the solid angle subtended at the eye, and p a function expressing the position of the source in relation to the direction of view.

For a number of sources, the effect is additive; i.e.

$$G = G_1 + G_2 + G_3 + \dots$$

Plate 2 Light sculpture at headquarters of Manila Electric Co.

Plates 3 & 4
Concealed sources in traditional interior and positive pattern of fittings in modern building

3

4

Plates 5 & 6
Austere furnishing, elaborate lighting: changing effects in show house

5

6

Plates 7 & 8
Low-brightness fluorescent fittings in industry and commerce

7

8

and, to give numbers that are easy to handle, the glare index is defined as

$$10 \log_{10} AG$$

where A is a constant, which was 1 when Imperial units were used.

The derivation of the index need not, however, concern the practising designer. He should be aware of the general way the glare effect changes with the variables on which it depends, and he should also appreciate broadly the significance of numerical values of the index. This understanding is best acquired by checking the index of schemes he encounters in practice, against the recommendations for limiting values given in the IES Code. Typical entries include

fine processes in watchmaking	10
hospital wards	13
drawing offices	16
university laboratories	19
circulation areas	22
hotel foodstore	25
iron and steel works	28

A change of one unit in the glare index is scarcely noticeable; a difference of three—the interval in the recommended limiting values—is significant. The index can be calculated from the tables only for the conditions for which they were developed, namely a regular layout of one type of fitting.

Indirect or reflected glare is not easily subject to analysis, and is due as much to the finish of room surfaces and furnishing as to the lighting arrangement. But it is clear that, where a low glare index is called for in the direct effect of the lighting installation, we should give serious attention to reducing visual problems due to specular reflection in glossy surfaces. In some situations of fixed geometry, as in the relationship of pupils in a classroom to the chalkboard or of an industrial worker to his machine, it is possible to ensure that the images of bright sources occur where they cause no difficulty. Polarisation may significantly reduce some forms of reflected glare. Fittings with polarising diffusers have, however, not so far been used very extensively in Europe. The low transmission through the material, and the rather lifeless general effect in the interior, seem to outweight the advantages of the system, but this is a personal view and one that may need to be revised.

Over glare, in general, there are few absolutes; what would be intolerable in an office could be exciting in a fairground.

Chapter 4

Brightness balance

Looking at the brightest part of a field of view is a basic animal reaction. Making the visual task the brightest area one can see brings together conscious motivation and subconscious inclination. However, it is one thing to require that the 'job' itself be made brighter than alternative areas to which the attention may otherwise be diverted, and quite another to produce a numerical specification for the relative luminance of visual task, immediate surround, and background.

It used to be said that working in a room lit by a desk lamp alone was 'bad for the eyes', or in some other way unsatisfactorv, because the contrast between thc work and the surroundings was excessive. A figure of 10 was quoted, as a rule of thumb, for the maximum ratio of task illumination to the illumination of the rest of the room, the argument being that a higher contrast would imply a strain on the adaptation mechanism as one glanced away from the work and then back to it. It seems likely that, for any particular adaptation level, there is a range of brightnesses that permits useful seeing; but, whereas this may extend from 1/10th to 100 times the level to which the eye is adapted when the level is relatively low, it can extend from 1/100th to ten times when it is higher. The region where brightnesses are too high is obviously associated with glare, although, as we have seen, glare is influenced by more than brightness alone. If the conclusion, however, is that glare is more difficult to avoid as general levels rise, it certainly seems to be borne out in practice. We can also feel sure that any 10 : 1 rule needs qualification.

In principle, reflectance, ρ, is the ratio of the light reflected back from a surface to the light that was incident on it.

$$\rho = \frac{\text{reflected flux}}{\text{incident flux}}$$

For unit surface area,

$$\text{reflected flux/m}^2 = \rho \times \text{incident flux/m}^2$$

This equation may be expressed as

$$\text{luminance} = \text{reflectance} \times \text{illumination}$$

provided luminance is in reflected lumens per unit area, i.e. in apostilbs (asb). Unfortunately, this is not the preferred SI unit of luminance, but the theoretically simpler candela per square metre (cd/m^2) gives a more complicated relationship:

$$\text{luminance} = \frac{\rho \times \text{illumination}}{\pi}$$

Thus, if a surface of reflectance 50%, or 0·5, has an illumination of 800 lux,

$$\begin{aligned}\text{luminance} &= 0{\cdot}5 \times 800 \\ &= 400\,\text{asb}\end{aligned}$$

or

$$\begin{aligned}\text{luminance} &= \frac{0.5 \times 800}{\pi} \\ &= 400 \times 0{\cdot}318 \\ &= 127\,\text{cd/m}^2\end{aligned}$$

One should expect to find the perferred unit, candelas per square metre, used in formal statements, but the greater convenience of using apostilbs is likely to ensure their survival in practical design work. Some of the more important units and terminology of lighting design are described in the glossary at the end of this book.

The 1961 edition of the IES Code proposed 10 : 3 : 1 for the relative brightnesses of task, immediate surround, and background. Despite reservations expressed, these figures tended for a time to be regarded as some sort of magic formula for good lighting. The recommendation does not appear in the 1968 Code. In any case, the ratio of the luminances of the task and immediate surround must depend mainly on the reflectances unless a local lighting system of quite unlikely sophistication is employed. An illumination of 1000 lux on white paper with a reflectance of 0·75 resting on light grey blotting paper ($\rho = 0{\cdot}40$) implies luminances of 750 and 400 apostilbs; a medium grey desk top ($\rho = 0{\cdot}22$) could give 220 asb and so the specified 10 : 3 ratio (it may be a fallacy to think that much of the desk top is exposed under real working conditions, and the luminances around the immediate task area may depend quite fortuitously on the reflectances of the papers, file covers, and so on, that happen to be there). The required average luminance of 75 asb for the background might be produced by an almost infinite range of combinations of illumination and reflectance (150 lux and 0·5, 300 lux and 0·25, and so on).

Such thinking about the effect of lighting schemes was one factor that made the idea of 'luminance design' seem worth pursuing. A systematic approach based on a standard computation may be illustrated in specific numerical terms. Suppose we want to light a room so that the average luminances on the main surfaces are

ceiling	300 asb
walls	200 asb
working plane	400 asb

Since

$$\text{luminance} = \text{illumination} \times \text{reflectance}$$

$$\text{illumination} = \frac{\text{luminance}}{\text{reflectance}}$$

or

$$E = \frac{L}{\rho}$$

so if we take the reflectances of ceiling, walls and working plane as 0·7, 0·3, and 0·5:

$$E_c = \frac{300}{0 \cdot 7} = 430\,\text{lux}$$

$$E_w = \frac{200}{0 \cdot 3} = 670\,\text{lux}$$

$$E_{wp} = \frac{400}{0 \cdot 5} = 800\,\text{lux}$$

It is clear that a luminance of 300asb on the ceiling itself produces illumination on walls and working plane. The transfer of light from one surface to another depends on the geometry of the space, which can be expressed as a 'form factor', where

illumination on walls duc to luminance of ceiling = luminance of ceiling × appropriate form factor

Form factors are tabulated in Reference 2.16 and elsewhere. In the example we are considering, it may be found that a luminance of 300asb on the ceiling itself produces 150lux on the working plane, while a further 100lux results from the wall luminance. This means that 250lux of the required 800 on the working plane is produced by interreflection, leaving 550lux to be produced directly. In this way, the direct illumination on each surface is found, and multiplying it by the area of that surface shows what the direct flux from the lighting installation must be. The result is

flux to ceiling = 22000lm

flux to walls = 33000lm

flux to working plane = 45000lm

The installation must therefore produce 22000lm upwards and 78000lm downwards. As we shall see later (Chapter 12), this gives a 'flux fraction ratio' of 22000/78000, or 0·28. Further, of the downward light, 45000lm must be directly incident on the working plane—a 'direct ratio' of 45000/78000, or 0·58. These data reveal the essential nature of the fittings needed to produce a specified 'gross luminance pattern' in a room of known proportions and reflectances.

At first, this seems an elegant solution. Subsequent doubts are not about the answer so much as the question. There are, of course, a number of mathematical reservations and qualifications to be made, but the main difficulty is in formulating the luminance specification. Was there any basis for seeking to produce 300, 200 and 400 asb on ceiling, walls and working plane in the example above? Attempts at starting from subjective or apparent brightnesses and converting these to luminances have been no more than partially successful—and this is not surprising, because the link between the subjective impression and the objective measurement is the key problem in environmental studies generally. Until recently, it was assumed that apparent brightness depends only on luminance and adaptation. It was a moment in the tradition of 'the emperor's new clothes' when the few remaining innocents in the lighting world declared that they could, under normal circumstances, tell the difference between midgrey under a moderate illumination and dark grey under a higher level, even if the luminances were the same.

This brought lighting studies dramatically down to fundamentals, to what in many ways should be their starting-point, the consideration of how we see. The new-born infant has vision but cannot see. Vision is physiology, seeing is perception. Before we can proceed, therefore, we must look, even if very briefly, at some of the basic ideas that arise in the study of seeing in perceptual psychology (References 2.17, 2.18 and 2.19 are relevant).

The very young child experiences a confused barrage of sensations and makes no sense of them. It can survive only because it is protected by someone who has learnt to interpret sensory stimuli. In learning for himself the meaning of the messages his brain receives, the infant has to reconcile what he sees with what he touches. He may be born with an inherent awareness that there is a reality external to himself in which objects pursue an independent and basically constant existence—if not, he soon realises that this is so. Imagine a baby examining a rattle. He grows tired of it and throws it a little way off, then crawls over and picks it up again. During the initial examination, the image on the retina assumes a variety of shapes; when the rattle is thrown away the image becomes smaller. But when it is retrieved and examined again its unchanged character is confirmed. The baby is learning that the same object can appear to have different shapes according to its orientation, and different sizes according to its distance. For survival, recognition of the object is the vital achievement, and so the human being learns almost to ignore—or, in a sense, not to 'see'—the changes. The 'constancies' of size and shape are characteristic of normal visual perception.

In similar ways, we learn to recognise whiteness or lightness, even though grey under direct sunlight may have a much higher luminance than white in the shade of a tree, and to appreciate that the colour of a surface is unchanged even though its appearance under a north sky and lit by a tungsten lamp may be quite different. We speak of 'brightness constancy' and 'colour constancy', and we mean that in normal circumstances, the human being learns to ignore changes in appearance that he knows from experience are caused by particular lighting conditions. Again we might almost say that he does not 'see' these changes—unless they are extreme or untypical, or unless he is consciously looking for them because his attention has been alerted. Lighting engineers sometimes complain that laymen seem almost obstinately unaware of the effect of changes in the lighting of a room or a scene; the paradox is that it is just because they know from their visual experience how lighting tends to change appearances that they are able to see through it to what, without metaphysical quibbling, we can call the underlying reality.

Accurate perception depends on the conditions allowing the constancies to apply. In particular, brightness constancy is observed only when the range of luminances in the field of view is relatively restricted. When contrasts are considerable, we enter the sphere of illusion, drama and poetry; perception is no longer unambiguous. There are circumstances, of course, when this is desirable —in shop and exhibition display, in the theatre—but, for working interiors, the prose of lighting, accurate and immediate perception is the aim.

It is, then, in these terms, of the brightness relationships that permit constancy to be maintained and of the conditions under which it breaks down, that the study of desirable ranges of luminance is likely to progress.

Chapter 5

Colour

The subject of colour embraces physics, physiology and psychology; colour is an important part of man's culture generally and it is expressed in art, taste and fashion. Lighting is closely related to these aspects, and the question of colour arises frequently in later chapters. The immediate discussion is limited mainly to two topics, both influencing the choice of lamps for a lighting scheme: the colour rendering of light sources, and the psychological effect of warm and cool colours of light.

The apparent colour of a surface is to some extent in the eye of the beholder. Quite apart from the philosophical imponderables of such questions as whether the sensation you identify as 'yellow' is the same as mine, recent visual experience and fatigue have their effect—as do defects in vision. But the light flowing into the eye from a surface is a function of two things: the characteristics of that surface and of its illuminant. Apart from white, black and genuinely neutral grey, all surfaces are differential reflectors. Suppose we take the visible spectrum as made up of just four 'bands'—blue, green, yellow, red. A particular 'yellow' surface might reflect 10% of the blue falling on it, 30% of the green, 70% of the yellow, and 30% of the red. Suppose this surface is illuminated in turn by two sources, the first a theoretical 'white' light made up of 100 arbitrary units in each band, the second weak at the ends of the spectrum and composed of blue 50, green 150, yellow 150, red 50. In the diagrams in Fig. 1.6, the top shape shows, in these rather crude terms, the spectral composition of the incident light. Below it is a representation of the differential reflectance of the surface, and at the bottom the resulting spectral composition of the reflected light. As we should expect, the light rich in yellow and short of red has the effect of emphasising the yellowness of the surface while giving it a slight green tinge. Numerically, the height within any band in the bottom diagram, i.e. the band content in the light reflected from the surface, is obtained by multiplying the height in the top section (the band content in the incident light) by the reflectance of the surface in that band—or 'top' times 'middle' equals 'bottom'.

The diagrams illustrate the dependence of the apparent colour on both surface and illuminant. It is perhaps tempting to think that they also show that only 'white' light can reveal an object in its 'true' colours. Although this might be one way of defining the 'true' colour of a surface, it is not a very useful idea, since white light of this kind—which has a horizontal straightline as the graph of its spectral composition—is something we do not normally meet from either natural or artificial sources. We make a major advance in understanding the relationship of colour and lighting when we recognise the fact that sources with very different spectral compositions may all produce effects we accept as 'natural'—while we realise, of course, that there are others which do not.

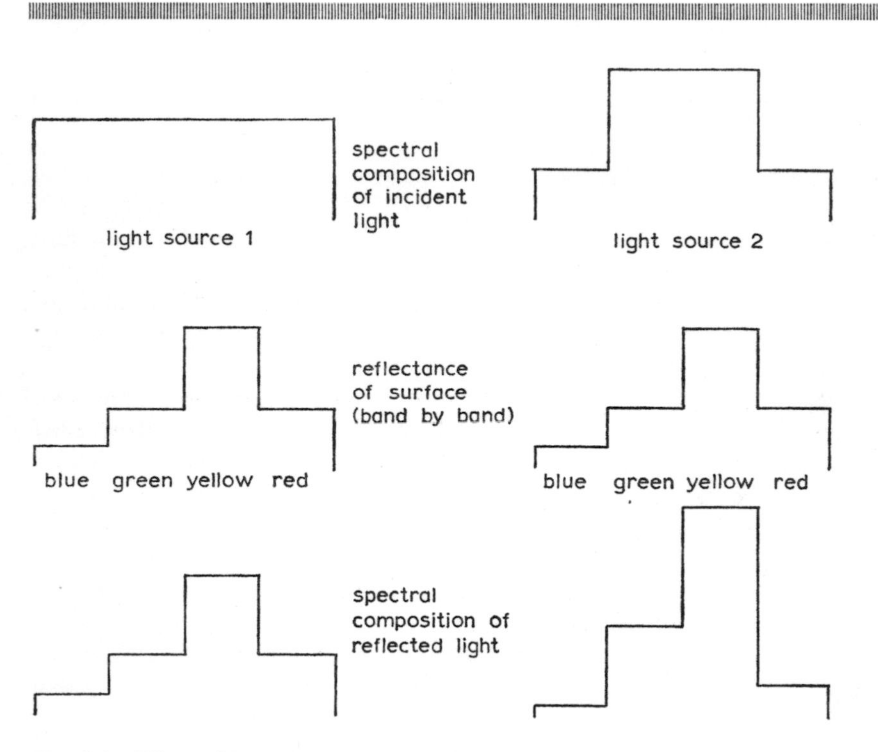

Fig. 1.6 Effect of light source and surface characteristics on colour rendering

Sources with a natural effect include a blue sky, a cloudy day, direct sunlight, some fluorescent lamps, incandescent lamps, candles. . . . Sources that produce some sort of distortion, an unnatural effect, include many discharge lamps and some fluorescent tubes—and the sky under some stormy conditions. What quality do those in the first group share which is absent in the second? (The distinction may be less absolute and more of degree than this wording suggests, but it is useful to simplify at this stage.)

If a lump of solid matter is heated, it eventually reaches a temperature at which it emits light—it becomes incandescent. Historically, nearly all 'artificial' light, from the fire in the cave to the filament lamp, depends on incandescence. In a luminous flame, the particles of carbon, or other solid matter, within the flame are heated by it and so give out light; in a normal household bulb, the passage of current through the resistance represented by the filament raises it to an appropriate temperature. The emission of light from an incandescent body increases rapidly as its temperature rises, and the colour of the source changes from an initial dull red to a whiter appearance.

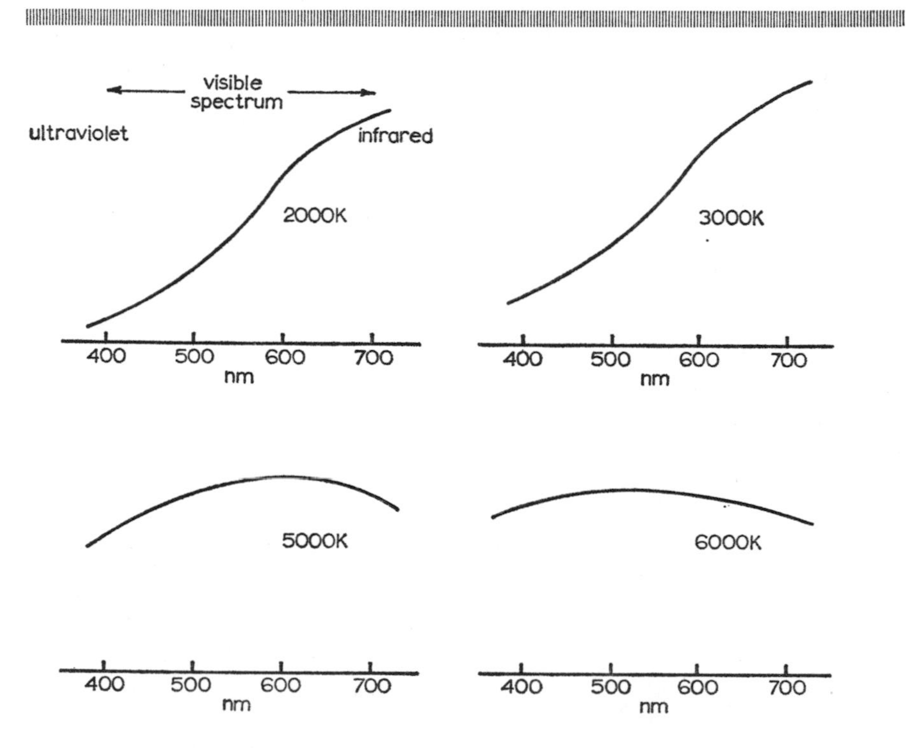

Fig. 1.7 Relative distribution of energy within visible spectrum in radiation from 'black body' at progressively high temperatures (not drawn to same vertical scales)

A 'full radiator' or 'black body' has a surface that emits the maximum radiation possible at any particular temperature. Many real surfaces correspond fairly closely, in the way they radiate, to the behaviour of this theoretical 'ideal' surface. The graphs in Fig. 1.7 show approximately how the composition of the visible radiation changes as the temperature rises (the lines have been drawn separately, since the total quantity of radiation increases so rapidly with temperature that difficulties of scale appear if the attempt is made to show them on the same graph). The progressive change from 2000K is one from light that has little blue and is relatively rich in red to radiation at 5000 or 6000K that has much more nearly equal contribution from each band of wavelengths in the visible spectrum (the graphs are, however, still curves).

If the appearance of a real light source matches that which a black body will have at a particular temperature, say 4560 K, we can describe how the real source looks by saying that its 'colour temperature' is 4560 K. But, since many different mixtures can produce the same total colour appearance, we cannot conclude that the spectrum of the source has the same relative composition as a black body at that temperature; this is an additional piece of information that may or may not apply. It is, of course, on this spectral composition that the colour rendering depends. The point—an important one—is that two sources may appear to match, but produce quite different colour effects.

All this may seem theoretical and with little immediate relevance to lighting practice, but, when we consider a wide range of sources that produce a natural colour effect, we find not only that their appearance can be matched to a black body—so that it can be described in colour-temperature terms—but also, and significantly, that their spectral compositions do agree closely with the corresponding full radiator:

blue sky	10000–20000 K
overcast sky	6000–7000 K
noon sunlight	5000 K
afternoon sunlight	4000 K
incandescent lamp	2800–3200 K
candle flame	2000–2500 K

On the other hand, a lamp like the Warm White fluorescent tube can reasonably be assigned a colour temperature in the 3000 K region, but its spectrum has much more green and yellow, and much less red, than a full radiator at 3000 K—which accounts for the distortion we recognise in its effect on coloured materials.

There is, then, no single source of light representing an absolute standard of colour-rendering excellence (although standard illuminants may be defined for reference purposes). The criterion of colour-rendering quality normally accepted is the closeness with which the spectrum corresponds to that from a black body of the same colour temperature. The composition of the spectrum can be analysed in terms of the content of a number of bands of wavelengths (six and eight bands are used in two such systems) and a figure of merit or a classification declared as a result.

Another approach to the problem of awarding a 'mark' for colour rendering is to base it on measurements of the shift in appearance of colour samples under a standard illuminant and under the source in question. The Crawford classification, based on work at the National Physical Laboratory, rests on spectrum-band analysis; fluorescent lamps, for example, can be categorised A,B,C etc. The 'colour shift' approach was used in the development of the colour-rendering index recommended by the CIE (Commission Internationale de l'Éclairage, the international lighting body which operates in parallel to national lighting institutions such as the UK Illuminating Engineering Society).

A lamp with perfect colour rendering would have a CIE index of 100—though 4 or 5 units is the smallest difference a typical observer may detect. Most fluorescent lamps have an index between 55 and 95. The practising designer need not be familiar with the detail of the theoretical basis of colour-rendering indexes, but he does need to know of their existence and to have some ideas of the comparative quality of, say, Crawford's B and E categories. Lighting recommendation for interiors are beginning to include colour-rendering specifications as well as those relating to illumination levels and glare.

The second topic that merits discussion here, following these ideas on colour rendering, is the psychological significance of warm and cool light.

Primitive man again provides a pointer. He is at his most active at noon, when exterior daylight is usually at its strongest, and most blue. He continues with his hunting through the afternoon, while typically the light is getting less intense and its colour temperature is falling; i.e. in the way we normally use the words, the light is becoming warmer. As the sun sets, the light from the sky is rich in red, and, when it fades altogether, the hunter retires to the fire in the cave with its low illumination and all that it means: food, security, relaxation.

Although there has been little formal research to confirm the idea, we can reasonably assume a basic association of high illuminations with cool light and with work. Correspondingly, it seems essentially appropriate that evening and social activity should go with lower illumination from warmer sources. Climate, of course, affects expectation and reaction; in a hot country, we may not feel the need after dark for the cosy associations of light in the 2000–3000 K colour-temperature range, but in the absence of daylight, lighting levels are without an obvious external standard of reference, and the resulting low illumination tends to make warmer light seem appropriate. The subject is one we return to when we look more comprehensively at the relationship of daylight and electric light, and when we come to interiors, such as art galleries, where colour is particularly significant. The idea that basic reactions to warm and cool light might be common to humanity seems the more plausible after reading a hypothesis put forward by René Dubos in 'Man, medicine and environment':

> 'The brain of modern man and his fundamental mental processes were certainly moulded early during the Ice Age: the formative influences derived from the interplay between the culminating phases of his biological evolution and the initial phases of his cultural development. These influences constitute, therefore, the common background out of which emerged the fundamental genetic makeup and cultural traditions of mankind. Secondary adaptations producing the various human races occurred only after the basic formative processes of anatomical, neural and cultural development had been completed.'

No note on colour and vision would be complete without at least a brief mention of three related phenomena: colour fatigue, colour adaptation, colour constancy.

Although colour vision is not completely understood, it is thought that the nerve endings in the retina include three types of cone sensitive, respectively, to the primaries green, blue and red. If the eye receives light which is predominantly blue, say, the blue-sensitive receptors 'work harder' than the rest, and become fatigued; so that, if subsequently the gaze is directed at a white surface (emitting comparable quantities of green, blue and red), the depression of the blue response seems to exaggerate the response to green and red. The impression is yellow—which may be regarded as a mixture of green and red, or the complementary of blue, in other words white minus blue. Most coloured after-images owe their existence to this phenomenon of colour fatigue.

It also explains what happens if circumstances are contrived to allow us to see similar objects and surfaces simultaneously under, say, a north sky and tungsten lighting. While we expect a difference in effect, its extent surprises us. This is because we are normally within a completely daylit or tungsten-lit environment. Colour fatigue tends to compensate for bias in the illuminant by rendering the eye less sensitive to whichever wavelengths are present to excess. This colour adaptation is, however, never complete—if it were, any old lamp would do.

The terms 'colour fatigue' and 'colour adaptation' tend to be applied to physiological mechanisms. There is also the perceptual aspect. One part of learning to see is learning, from experience, how to disregard the effect of particular lighting conditions; and these conditions include the colour effect of the source. It has been said that we see what we expect to see, and most people confronted with a London bus under nothing but sodium light would 'see' it as red, even though there is no red present in the incident light to be reflected to the eye. Only if their attention were especially alerted would they realize their 'mistake'. Take another example: imagine a groundfloor room glazed along one side with sunlit grass beyond; the wall opposite the windows is painted red, and both crosswalls are white. If we are to examine one of the crosswalls carefully, we shall observe that it has a definitely greenish tinge at one end owing to reflection from the grass, and that it is pink at the other where light from the back wall bounces on to it; but, in the ordinary way, we shall probably not notice these effects, and see it simply as a white wall. As the man said, what is truth?

Colour constancy is one of four constancies (with those of size, shape and brightness) that are characteristic of visual perception; they are essential to us if we are to make sense of our surroundings.

Chapter 6

Modelling

Imagine a featureless snow-covered landscape under an overcast sky. This represents probably the most diffuse lighting encountered in nature, and recognition of objects under these conditions depends mainly on contrasts of reflectance and colour, and on extrapolation from our more normal visual experience.

Think now of the effect of a single concentrated spotlight in a room lined with black velvet. This is clearly the other extreme, directional lighting at its most exaggerated, with contrasts between highlight and shadow greatly exceeding those normally occurring in nature.

The balance between diffuse and directional light is a very significant element in total lighting effect. If we do not achieve a balance of this kind within the range appropriate for the situation, a lighting scheme will fail, however suitable its illumination, glare control and colour characteristics. Office lighting is a case in point. Many routine flourescent schemes consisting of nearly continuous rows of opal plastics diffusers, in rooms of all-over high reflectance, produce interiors flushed with characterless light, basically dull, however great the illuminance. On the other hand, in some offices where the management has tried too hard to be trendy—in an advertising agency, perhaps, or a superior turf accountant's—spotlights alone produce excessive contrast and diversity.

On what we normally consider to be a fine day, the visually sympathetic impression is created by direct sunlight from a blue sky. This represents some sort of standard for diffuse and directional light, not because the particular balance will be appropriate for more than a limited range of situations, but because it provides a basis for comparison.

The term 'a blue sky' covers a wide range of conditions, depending on latitude and the amount of water vapour in the atmosphere. The deepest blue occurs high in the dome, but away from the Sun, with a much whiter colour near the horizon, particularly when there is a slight haze. Nevertheless, the total effect of the sky is of very diffuse and markedly cool light producing perhaps 5000–10000lux on the horizontal ground. The Sun gives a strong, parallel beam of warmer light (the colour temperatures might be 10000K and 5000K) producing as much as 50000lux on a plane perpendicular to the flow. The ratio of the illuminations in highlight and shadow is thus typically between 5 : 1 and 10 : 1 with a definite colour contrast. One lesson may be drawn at once: directional light is more likely to produce a broadly natural effect if it is warmer than the general diffuse illumination—and, of course, if it shows some coherence in direction.

What should be noted, however, is that it is not necessary for the colour temperatures to be those occurring in nature; indeed, if they are, something approaching out-of-door illumination levels will be necessary for a comparable subjective effect. The principle encountered earlier, that high illuminations and high colour temperatures go together, might be developed into the hypothesis that, as illumination is reduced, so the colour temperature must come down, in some sort of scale, for the same subjective impression to result. A precise numerical law would be difficult to formulate, but many observers would agree that tungsten directional light (3000K) against a background of modest interior illumination from 4000K tubes would produce an impression not unlike that of direct sunshine (5000K) against a background from a 10000K sky at exterior levels.

It is a fallacy to assume that directional light necessarily means parallel beams or point sources. The characteristic quality of light caught, for instance, in many paintings of Dutch interiors is that of flow from a window, i.e. from that part of the sky effectively visible through it. Light within a room with one window, or windows confined to one wall, has a unity and clarity of effect that is subconsciously appealing.

Traditionally, architects are advised to include windows in more than one wall of a room if possible, so that light from the second should reduce the contrast of brightness around the first; this becomes more difficult as buildings get deeper, but the gain in coherence is at least a compensation. For some years now, lighting designers have pursued the idea of producing a flow of light across a space to give a degree of directional interest to the interior. In a sales area in a shop or showroom, with general fluorescent plus tungsten spotlighting, the beams of the latter may be directed with a basic axis in mind; similarly in a series of single-aspect display windows along the front of a large department store, the main flow of directional light can be intentionally one way or the other. In working interiors, and offices in particular, the tendency has been to explore the effect of fluorescent fittings with an asymmetric light distribution. In a small office, with glazing along one side, fittings of this kind can produce a flow of light away from the window wall, echoing the spread of daylight. But even here, when one looks towards the window, i.e. against the flow, the increased chance of glare is obvious. The situation may be acceptable if the desks are arranged so that the usual direction of view is parallel to the outside wall. In a large room, the glare potential increases, and there is less likelihood of a restriction on desk arrangement being acceptable, since extensive open offices have, as part of their justification, complete flexibility in the grouping and regrouping of furniture.

It is, incidentally, one of the many enigmas remaining in lighting that it seems easier with natural light to reconcile a directional flow and subjective freedom from glare. Replace a real side window with an artificial one of precisely the same area and brightness, and most people will declare it more glaring; perhaps the distance of the sky, or even its lack of a clearly defined distance, has some effect. At any rate, one of the few electric-lighting techniques you can at the moment pursue with confidence, where the intention is to produce a sense of flow across a large working space, is that of developing an acceptable scheme with symmetrical light distribution, and then restricting the flow in one direction so as to create, as it were by subtraction, an emphasis in the opposite. Even with this approach, of course, visual comfort will vary with viewpoint.

Completely different techniques apply, naturally, to photographic, television and stage lighting. The newsreader—provided that he can read the teleprompter—has no more justification in complaining of glare than does the actor. His head becomes 'the head', an object to be lit for its best effect on the spectator. The photographer's '3-point system' is essentially 'keylight', the major beam from one side at the front, 'fill-light', balancing softer illumination from the other, and 'backlight', a beam from above, behind and to one side to light the shoulders and produce a halo effect in the hair, mainly to separate the head from the background. Television frequently adds a second backlight, so that there is one from both sides, and a 'contrast control' source near the camera—hence '5-point system'. Knowledge of these techniques may be more important to the users of studios than to their designers, but the buildings that house these activities are naturally influenced by their practice. The terms 'key', 'fill', and 'back' light are basic in the language of lighting for effect, and are relevant to museum and art gallery lighting, and indeed to display generally. In addition, within many buildings, there are people with roles depending on face-to-face communication whose position may be known with enough accuracy to make possible their individual treatment—the receptionist at her desk or even the director at the boardroom table. Certainly, thinking of lighting in relation to the people in the building rather than to a series of empty interiors is a step in the right direction.

It is understandable that attempts should have been made to quantify the effect of directional lighting in a 'modelling index'. They have not so far been completely successful. The proposals by Waldram some time ago and Cuttle recently (Reference 2.20) seem too involved for routine application. A simpler index, but one limited in its application, emerged from the work of Lynes and others in research for Pilkingtons, the glass manufacturers (Reference 2.21). It depends on two concepts, themselves important in other ways. 'Scalar illumination' at a point is the mean spherical illumination, i.e. the average over the surface of a small imaginary sphere enclosing the point. The idea of 'illumination vector' involves imagining a plane through the point (such as a sheet of card) being turned in space until the greatest possible difference exists between the illumination on each side; then the numerical difference of the two levels is the magnitude of the illumination vector, and the line through the point perpendicular to the plane is its direction. Thus scalar illumination, as the term implies, represents the light arriving at the point irrespective of direction, and the illumination vector identifies the strongest flow of light through the point. The suggestion that the vector/scalar ratio is a useful index of strength of modelling has found some support from systematic comparison with subjective estimates, and recommendations of ranges of values for particular situations have been formulated in these terms. Not all those who attempt to use these ideas are as careful to acknowledge their limitations as those who proposed them—for instance, some nonsensical conclusions can follow the attempt to apply the illumination vector concept to the photographer's 3-point system, or indeed to any array of point sources.

Modelling may remain hard to quantify, but any lighting designer must attempt to concentrate his powers of observation on the ways form and texture are revealed visually, since these are at the heart of what we mean when we speak of lighting effect. A positive account of 3-dimensional reality has, moreover, a definite effect on mood and emotion. When Hamlet found all the uses of this world weary, stale, flat and unprofitable, the sky was surely overcast.

Plates 9 & 10 Relighting prompted by need to free space from intrusive hardware and to produce more positive directional effect. Beams from recessed incandescent fittings are inclined towards front of church

9

10

Plates 11, 12 & 13
Separate switching of upward and downward light gives valuable flexibility

12

13

Plates 14 & 15
Marked contrasts of brightness produce dramatic effects but may be perceptually confusing, more intensely lit areas seeming almost self luminous

14

15

Chapter 7

Variety

The terms 'variety' and 'flexibility' as applied to lighting overlap to some extent, and are often used loosely. Let us here look first at the variety that may exist within a particular lighting treatment of an interior, and then turn to the question of changes in that effect.

Traditional lighting practice developed during the years when the incandescent lamp was very nearly the only available electric source of light. The tungsten lamp being small, many early fittings were intended to spread its output as evenly as possible over a wide area—terms like 'dispersive' and 'general diffusing' occur frequently. Even with such fittings, there was a limit to how far apart they could be if acceptable uniformity of illumination was to be produced (it is, of course, not absolute separation that matters, but how spacing compares with mounting height). It was found that, for generally dispersive fittings, a limiting 'spacing/mounting-height ratio' of 1·5 was usually consistent with the minimum illumination at any one time not being less than 0·7 times the maximum (a '70% uniformity'). A fitting mounted on a ceiling 3·150m above the floor is 2·300m above the conventional working plane at a height of 850mm. For the stated condition to be observed, the layout must be based on a spacing not exceeding 1·5×2·300m, i.e. 3·450m. Approaching a routine lighting scheme in this way enables us to find the smallest number of lighting points consistent with acceptable diversity, and this is useful, since the fewer the points the more economic installation and maintenance are likely to be.

There is nothing sacred about the figure of 0·7, or 70%. Regarding this as a standard was just a convention—and one that applied only to working interiors —but it had a value that practice confirmed. Manufacturers usually state a limiting spacing/mounting-height ratio consistent with its observance. If not, the light distribution of the fitting is usually given in terms of a 'BZ number' (Chapter 12), and limiting ratios can be quoted in these terms:

BZ1 and BZ2, maximum spacing/mounting height ratio	1 : 1
BZ3 and BZ4,	1·25 : 1
BZ5 to BZ10,	1·5 : 1

This table must be applied with some caution, since the narrowest distribution category (BZ1) covers a fairly wide range; with some narrow-beam downlights, for example, a spacing only half the mounting height may be the limit if conventional uniformity is to be achieved; on the other hand, such fittings are unlikely to be used in situations where uniformity is critical.

However, the need for such limited diversity as that implied by 70% has been questioned recently. It has been argued that the convention became established when levels were very low by today's standards. While the average is, say, 80lux and peaks 100lux, it is understandable that it should seem desirable not to let the minimum fall below 70; but, when the average is 800lux, the extremes of 560 and 1120lux are perhaps acceptable. These apparently arbitrary figures are, in fact, 0·7 and 1·4 times the average, so that the ratio of maximum to minimum is simply 2. Such a ratio is noticeable in the lighted interior, but its advocates claim that this is positively welcome variety and that no one is going to complain that they have only 560lux—which seems to imply the principle that diversities greater than the conventional are acceptable so long as the minimum illumination significantly exceeds the current recommendation for the activity. Others argue that, with such a diversity in, say, a landscaped office, people tend to regard the areas under the fittings as privileged—as those near the windows once were—and this mitigates against the idea of flexibility. The issue, then, is unresolved. Perhaps the mistake is to seek a general rule-of-thumb solution to be applied to all cases. The mature designer will bear these factors in mind among the many that influence his ultimate decisions on layout.

The words we use are loaded, of course. Although diversity sounds like a necessary evil, variety is the spice of life. Uniformity may not seem terribly attractive, but giving everyone equally good conditions seems virtuous.

However, I would hope that, if anything has emerged from this discussion of criteria so far, it is that numerical standards should be regarded merely as guide lines. We may need to adopt them unquestioningly when time is pressing or the situation routine, but almost any piece of lighting design represents such a complex pattern of interrelated variables and values that it merits unique consideration.

Visual variety is, of course, more than a question of illumination levels over a working plane. The brightnesses presented to the eye by the whole visual field are part of it. In Chapter 4, we looked at the criteria of brightness balance, at the way task, immediate surround, and general background, might be related, and also at systematic luminance design, which sets out to produce a particular 'gross luminance pattern'. Variety also depends on contrasts and changes in brightness in detail, where the scale is small. The appeal of the traditional chandelier is in its scintillation. Although a lighting engineer tends to regard shiny finishes as regrettable, since they so aggravate the problem of reflected glare, strictly limited areas of gloss or chrome can vitalise an interior. The right word seems to be 'sparkle'—difficult to define accurately in this context but expressing a quality we often seek. Colours of high chroma and low value may need discretion in handling, but it would be a dull world if we were afraid to use them. Visual comfort need not mean visual boredom.

Chapter 8

Flexibility

Our increasing awareness of what every cubic metre of space within a building costs has led to a demand that it should be utilised to the full. At the same time, change has accelerated—in organisation and method in business, in production techniques in industry, and in conventions and customs socially. All this means that the interior fully designed for one specific activity must become less common. We may regret this, particularly if we have a developed interest in one particular use of the space—theatre people are always saddened when yet another council decides to erect a multipurpose hall—but the trend is inescapable. It is argued that modern technology ought to make it possible to adapt one interior for a wide variety of uses, and so it can, at a price. We have all met ambitious schemes for halls with dance floors that can be flooded for swimming or skating, tilted to provide raked seating, and raised to leave banqueting space underneath—mercifully, most have stayed on the drawing board once the operational and maintenance implications have been faced with realism.

However, visual appropriateness is an important part of what makes an interior suitable for a particular activity, and here changes in the lighting can produce the distinction we feel to be necessary, and do so with relative simplicity and economy.

Mechanical flexibility is, of course, important. In the simplest terms, it may be just moving the light source bodily, as when we unplug a table lamp and reconnect it in another room—not very different from taking the candle with you when you go to bed. A rise-and-fall pendant can have its position in space altered without disconnection, or a spotlight on a track can slide along to a new point. Many fittings producing a beam are adjustable in direction, and this facility means more than just changing the object being highlighted, since, in a room lit by a number of spots, the proportion of upward and downward light can be varied at will.

Some degree of mechanical flexibility is likely to remain useful, especially when initial cost must be kept low. In a university study bedroom, for instance, a fitting on a swinging arm can serve both desk and bedhead. But the underlying movement in technology is towards the replacement of mechanical change by electrical or electronic. The semaphore direction indicator on cars gives way to the flashing type, the acoustic gramophone disappears, the abacus becomes the computer. In lighting, the single operating-table fitting on its intricate mounting is replaced by a device allowing the surgeon to choose one or two spotlights from many recessed over his head; or, in the theatre, the follow spot, needing a man to direct it, is superseded by sequential dimming in and out of a row of lanterns under remote control by the single board operator with his sophisticated electronic equipment.

Simple switching is the primary means of electrical control of lighting—so obvious a statment that one would hesitate to make it, were its implications not forgotten so frequently. Take a combined living room with a ceiling fitting in the dining area and another in the sitting part. If they are on one circuit, all you can do is to have light or not. But if they are switched separately, you can have either, or both, or none—that is, three positive alternatives. Where there are n circuits in an interior, the number of different lighting effects made possible just by switching is 2^n-1 (each circuit may be on or off, but we are ruling out the case where they are all off). Thus with, say, four circuits, there are 15 different possibilities.

What we find, of course, is that, although the search for this sort of flexibility may be prompted originally by an interior having a number of different uses, we enjoy the variety of effect in itself and want to exploit similar freedom everywhere; and, after all, any space in a building, however modest, is used at different times of the day, on different days of the week, in spring and in winter, by old and young, happy and sad. I believe that what most people object to visually in having to spend any length of time in an artificial environment is the unrelenting sameness of the lighting, good or bad. Natural light is always changing, even if at times almost imperceptibly, and this is one of its important characteristics. As buildings become deeper, we need to identify which aspects of natural conditions are most valued, consciously or subconsciously, by the human being, so that we can provide not necessarily an imitiation but some compensating feature.

It is the timing, the rhythm of natural change, that seems to matter most. A man's mortality is important to him; he does not inhabit eternity; and the events that punctuate time confirm the pattern of his existence. Think of all the significance, associations and symbolism that attach, for example, to morning. I remember, as a child visiting an aunt, a picture called 'The Bitterness of Dawn': a Regency gambler, having lost his all, sits disconsolate at the table his companions have left as the thin grey light filters past the curtains. The silent film's most quoted title was 'Came the dawn . . .'. Or in Hamlet again, after the apparitions of the night, there is a complete change of mood with Horatio's

But, look, the morn, in russet mantle clad
Walks o'er the dew of yon high eastern hill

—a lighting cue if ever there was. René Dubos has summed it up in the book mentioned earlier, 'Man, medicine and environment':

'Many important physiological functions of man exhibit diurnal, seasonal

> and lunar rhythms that persist even when a person is so completely shielded from changes in temperature and light that he is not aware of the movements of celestial bodies or of the passage of time. These rhythms were inscribed in man's genetic makeup at a time in evolutionary development when human life was closely linked to natural phenomena. The movements of the earth around the sun and of the moon around the earth are cosmic events unchanged since man acquired his fundamental characteristics. Even in his present artificial environment, biological rhythms persist in modern man. He may be unaware intellectually and socially of the diurnal, lunar, and seasonal influences affecting his body, but he cannot escape the physiological and mental consequences of these cosmic forces.'

Dimming adds enormously to the range of effects possible from quite modest installations, and manual control is all that is needed in most cases. The electrical control of light output is simpler with incandescent sources than with discharge types, and, in a mixed installation, it is often effective to dim the tungsten lamps while switching the others. A recent project for an amenity centre as the focus of community activities in a medium-sized town includes a building labelled simply 'covered space', the intention being not to limit its use by preconceptions. The lighting scheme proposed depends on upward lighting from both mercury and high-pressure sodium lamps and downward illumination using tungsten–halogen lamps. The roof, and the light reflected from it, may thus be bluish white, golden white or a mixture—or it may be unlit, and, for each of these situations, the dimmers conrolling the direct light can be at any setting. The exposed roof structure carries in addition lengths of lighting track, so that further fittings may be added, to give emphasis to a platform for instance. Generally, the separation of the upward light and downward light adds to the ease of creating a variety of effects in an interior, particularly if one component is on dimmers.

A great deal of interest has been excited in the last few years, and not surprisingly, but the more extreme forms of kinetic lighting, in which effect follows rapidly on effect, either according to an automatic programme or in response to some stimulus such as sound. Dramatic displays of this kind are usually the concern more of the users of the building than of its designer, the only demand on him being for an appropriate power supply. Yet these are developments that we should watch with interest, since the techniques that they explore can find application in the more subtle changes of effect that seem certain to be used increasingly in normal building lighting.

Changes in electric lighting within the hours of daylight and outside them are discussed in Chapter 14. The need for such differences is now widely acknowledged, but, in a sense, they represent only a special case. In a no-daylight factory, for instance, we might also specify a difference between morning and afternoon. This could be quite limited—a change in the relative brightness of the facing perimeter walls perhaps. And the argument for it would not be invalidated if it emerged that most people were not consciously aware of it. This could, in fact, be an advantage in a working interior—but it is too early to guess. What we need is the experience of applying these ideas. It is certain that there is more to lighting than the limited criteria of traditional illuminating engineering. It was encouraging to find a paper exploring many aspects of 'Design for variety in lighting' being presented to the IES national lighting conference in 1970 (Reference 2.22).

Chapter 9

Cost

Finding out what lighting will cost is not easy, and it is as well to admit this at the start. There are so many elements in any calculation that are at best estimates or predictions that the precision possible is limited, and the use of sophisticated techniques therefore rather doubtful.

Let us look first at a simple attempt to compare two supposedly equivalent schemes for providing 150 lux in a room measuring 6 m × 5 m. We can then consider the shortcomings of the approach and possible ways of refining it.

With a conventional ceiling height, light surface finishes, and ceiling-mounted fittings of reasonable quality, the required level of 150 lux might be achieved with two alternative schemes (Fig. 1.8):

(*A*) 4 fluorescent fittings, each for 1 × 65 W Natural

(*B*) 6 tungsten fittings, each for 1 × 150 W g.l.s.

Traditionally, costs are separated into initial and running costs, the former covering the equipment and its installation, the latter energy costs and maintenance (including cleaning and replacement of components with finite lives). Let us assume that the fittings cost £10 and £2 for the fluorescent and tungsten types, respectively, and that installation is £5 per point. Then the initial costs may be set out as

Fluorescent	Tungsten
4(10 + 5) = £60	6(2 + 5) = £42

Cleaning may be carried out at regular intervals, however much or little the lighting is on, but energy consumption and lamp replacement depend on the annual hours of use. Take this as 1000.

Consider energy costs first. The total load of the fluorescent scheme (including control gear) is 320 W and of the tungsten 900 W; so the annual electricity consumption figures are 320 kWh and 900 kWh. At, say, 1p per unit, this means

Fluorescent	Tungsten
£3·20 per year	£9·00 per year

Now, maintenance. Conventionally, the life of a lighting scheme is taken as

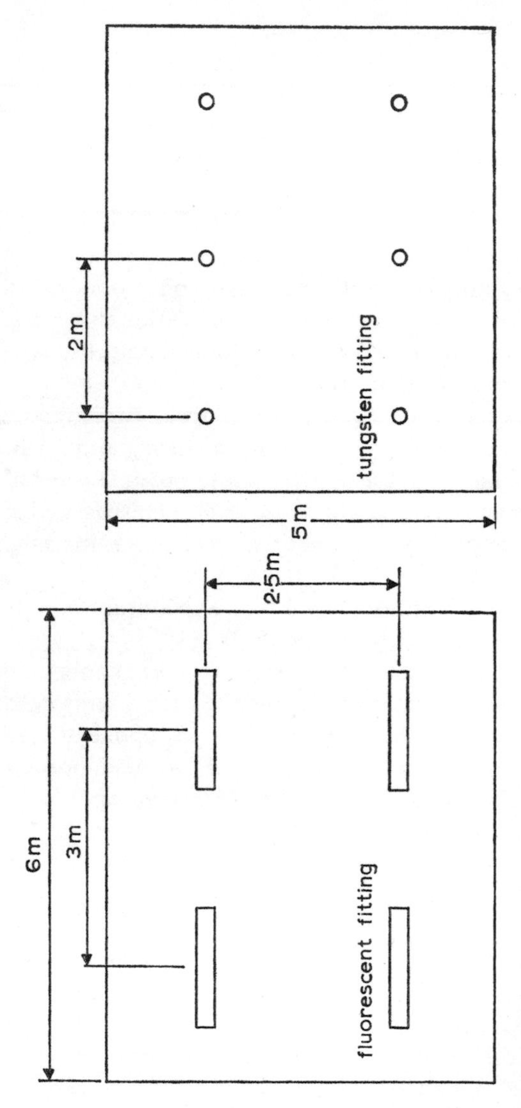

Fig. 1.8 Alternative lighting schemes

Time, years	Fluorescent scheme	Tungsten scheme
0	fit initial tube	fit initial lamp
½	clean fitting	clean fitting
1	clean fitting	clean fitting and replace lamp
1½	clean fitting	clean fitting
.	.	.
.	.	.
.	.	.
.	.	.
5	clean fitting and replace tube	clean fitting and replace lamp
5½	clean fitting	clean fitting
.	.	.
.	.	.
.	.	.
.	.	.
9	clean fitting	clean fitting and replace lamp
9½	clean fitting	clean fitting

ten years. This means that ten tungsten lamps will be needed for each fitting (the initial lamp and nine replacements), whereas, for each fluorescent fitting, there is the initial tube and one replacement. Let the pattern be as set out above: Since the labour cost of putting in the first lamp can be expected to be included in the installation cost, this lamp can be considered with equal justification as first cost, but, any in case, the sums involved are, say,

Fluorescent	Tungsten
4 × 70p = £2·80	6 × 14p = 84p

After this, the cost for labour and materials are

	Flourescent	Tungsten
Cleaning	18 times at 40p per fitting	10 times at 20p per fitting
Cleaning and relamping	once at 140p per fitting	9 times at 50p per fitting

Thus the total costs for ten years' operation may together be

	Fluorescent	Tungsten
Fittings and installation	60·00	42·00
Initial lamps	2·80	0·84
Energy	32·00	90·00
Cleaning	20·80	12·00
Cleaning and lamp replacement	5·60	27·00
	£121·20	£171·84

Although this calculation is open to many criticisms—and we shall look at some of them—it shows features that seem to apply generally and are worth comment.

The largest item is the energy cost for the tungsten scheme, and the next largest is the initial cost for the fluorescent scheme; maintenance costs are of the same order in each case; and, with about three hours' use of the lighting per day, the fluorescent scheme is significantly more economical over the life of the installation. It is generally true that fluorescent schemes are more expensive to put in than tungsten but cheaper to run (unless the hours of use are very short). Since fluorescent tubes cost something like five times as much as tungsten lamps, but last longer in roughly the same proportion, lamp costs are often comparable. More labour is involved in cleaning fluorescent fittings and probably lamp replacement, but the latter operation is less frequent. So total maintenance costs are broadly similar. In a typical scheme in an office used throughout the day, energy costs probably represent 60–70% of the total, maintenance 15–25%, and capital investment 15–20%.

Let us turn now to some of the many reservations that must be made.

First, the schemes may be equivalent in terms of illumination on the working plane, but they are so in little else. There are differences in colour rendering and colour temperature, in heat load, in light distribution and in the general visual impression. Even though the chance of trouble from noise, flicker, or radio interference is small with the fluorescent scheme, it is completely absent in the other case. If we had chosen high-efficacy tubes, such as Daylight, the economic advantage would have been more pronounced, but the visual differences greater.

Secondly, many of the figures, confidently fed into the calculation, are arbitrary. This difficulty in finding reliable data applies particularly to the labour content of maintenance, but it shows itself in many other ways. The £10 taken for the fluorescent fitting is reasonably typical of a medium cross-section batten and a reeded diffuser, but a single 65 W fitting may cost anything from £5 to £15. The range for 150 W tungsten lamps is probably even wider. Installation costs vary enormously, if only in their expression, because of the difficulty of defining where the lighting part of the total electrical services begins; perhaps, in a comparative exercise, one should try to estimate merely any marked differential.

Then energy charges are rarely simply a flat rate per unit. Most tariffs are in two parts: a 'maximum-demand' charge based on the greatest load, and a unit charge. The effective cost per unit thus depends on the hours of use. Since lamp replacement also depends on how much the lighting is utilised, it becomes clear that, where it is difficult to estimate this with any confidence—and this is the case in a high proportion of schemes—the validity of any detailed calculation is doubtful.

A different sort of objection may be made to the practice of simply adding up the total cost over the assumed life of the installation. This is that it makes no distinction between a pound spent now and one spent in several years' time. A pound five years hence has a 'present value' equal to the money that would need to be invested now to produce one pound in five years at some appropriate rate of interest. It would thus be possible to rework the calculation so that every cost, whenever it occurred, was rewritten as its present value; these figures could then be summed. In the case examined above, the major item in the fluorescent-scheme costs—for fittings and installation—occurs now, whereas, for the tungsten alternative, the largest figure, for energy, is spread over ten years, so that its present value is less than the total of the electricity bills. It can be argued that simply adding up the uncorrected costs exaggerates the advantage the fluorescent scheme shows. However, inflation has the opposite effect. Although we would prefer to pay a pound in five years' time to paying it now, we could find it cheaper to settle at once rather than face what the cost of the material or service would then have become. These two effects work in opposing directions, and, although it may be optimistic to assume that they exactly cancel each other out, they do mean that the error in just adding the totals calculated at present is likely to be less than that arising from all the other guesswork.

An alternative approach to costing in terms of orthodox building economics is to calculate the equivalent annual cost in use, that is, in effect, to add to the actual or predicted running costs (energy and maintenance) an amortisation of the first cost over the anticipation life of the installation. This gives results close to those from the 'present value' approach, and suffers from the same limitation of depending on estimates of interest rates and price rises.

My conclusion is that there are so many unpredictables in lighting costs that simple methods are likely to be as 'true' as the more elaborate methods. They have the advantage not only of being easier and quicker to apply but also of making it more obvious that they are limited in their accuracy.

Circumstances can arise to make prediction more confident. Suppose, for instance, a factory is to be expanded to give an additional production area equal to the area in existence. If detailed records of maintenance costs have been kept over the years, the works engineer will find them useful in evaluating the cost implications of proposals for the new building. Such closely relevant data are, however, rarely available.

In any case, it may be idealistic to imagine that designers are frequently being asked to consider the comparative economics of alternative proposals, however desirable this may be. Schemes are developed against a broad background of knowledge of what is likely to prove more expensive and what less; some sort of costing, usually just of the equipment, is worked out and passed to the quantity surveyer. The lighting designer may find himself having to justify a higher figure than had been anticipated (and he may bring running costs into this discussion), or he may be faced with the need to make 'economies', i.e. some redesigning. Even on this modest level, problems remain, among them finding out what equipment is really going to cost. A formal quotation may mean delay, or it may need a detailed specification that is not convenient at the time. Price lists are far from comprehensive and soon out of date, and tend to present a confused picture of retail, trade, users', architects' and other prices further muddled by fixer's discounts, prices 'on application', and the rest.

We have to live with a situation that is unsatisfactory. The plain fact is that, at the design stage, the costing of lighting schemes can be no more than a guide for an approximate budget.

To conclude on a slightly more positive note, we might list some rules of thumb (subject to exceptions and contradictions as they always are) for the achievement of economical lighting schemes; the order is random:

1. Assume that fluorescent lamps are the norm, and tungsten needs justification.
2. While regarding de luxe tube colours as standard, and recognising the fact that special types are sometimes necessary, always consider whether high-efficacy tubes might not be acceptable.
3. With tungsten, regard general lighting service lamps as the likely choice; reflector or other special types need justification.
4. Reduce the variety of lamps and fittings within a building as much as possible.
5. Do not use a sophisticated fitting where a simple one will do just as well.
6. Use the smallest number of lighting points consistent with the design criteria (in other words, do not use many low-wattage lamps where a few powerful sources are acceptable).
7. Consider whether devices for extending lamp life are justified.
8. Remember that maintenance can easily represent one-quarter of the total cost of the lighting; a scheme that is difficult, and so expensive, to maintain is unlikely to be economical.

The list can be extended, but the last point leads us happily to the next topic.

Chapter 10

Maintenance

The aim in most interior lighting design is to satisfy the demands implied by the visual criteria that have been discussed, and to do so—as Chapter 1 suggested—'at an acceptable cost in a way that permits effective and economic maintenance'.

Some aspects of cost have just been examined, and it is clear that maintenance is a significant element in the lighting budget. Recognition of its importance has grown slowly, and it is only recently that the need for a systematic approach to lighting maintenance has been accepted generally. Procedures and techniques are described in the IES report on depreciation and maintenance of interior lighting, which covers daylight as well as electric systems (Reference 2.9). The lighting designer can only benefit from knowing how fittings are cleaned, how group lamp-replacement schedules are worked out, and so on; but our immediate purpose here is to look at some of the direct ways in which maintenance considerations can affect the design of a lighting scheme.

The first, basic principle is that there must be access to the equipment; but, as soon as we begin to think about this, we realise that it is a matter of degree. Access is always possible, even if it means scaffolding or a helicopter. What is needed is a reasonable relationship between the ease or difficulty of access and the frequency at which attention is required. Thus tungsten lamps, with their short lives, are normally acceptable only if they can be changed easily; conversely, where access is inescapably awkward, a source with a long life is probably essential. Spelt out like this, the idea seems painfully obvious, but time and time again one encounters tungsten downlights, for instance, in very high ceilings. This could be legitimate if the importance of the effect justifies having a tower wagon handy, but such implications must have been considered at the design stage—otherwise we shall find a flight of steps preventing us from getting the tower wagon into the interior.

Immediate, permanent access to the equipment in position is of great value. Where ceilings have walk or crawl space over them, maintenance from above has many advantages in general, and particular conditions can give it extra point. In a swimming pool, for instance, it means not only that there is no problem in getting at fittings over the water, but also that the light can come through portholes in a sealed ceiling. The value of this is that the lighting equipment is away from the corrosion hazard due to the chlorine treatment of the water. Operating theatres or areas in industry with potentially explosive atmospheres are further examples of interiors that benefit from the removal of the electric-lighting equipment to another space. In almost any large interior, the cost of providing access from above is worth considering, particularly in auditoria where a raked floor and permanent seating complicate the approach from beneath. In a small theatre with a studio or workshop atmosphere, a series of exposed bridges can be invaluable for the stage lighting, and can often at the same time give access to the house lights. An alternative is to produce a ceiling from a series of panels on raising and lowering gear; the stage lights and the winches can be operated from the fixed bridges, which the suspended panels largely conceal, and lowering the panels makes maintenance of the house lights possible at floor level.

A number of firms make raising and lowering gear for lighting equipment. One thinks of it in relation to the traditional chandelier-type fitting, but the system has a much wider application, including recessed fluorescent types if necessary.

Another device from a number of manufacturers is the pole-type lamp changer, a long (sometimes extending) rod with a mechanical grab for seizing the lamp firmly. Though considerable feats of virtuosity are possible, delicacy and practice are needed, particularly as the height increases; so, though this may be the answer to an existing problem, I feel that it should not excuse the designer from providing some other way of changing lamps in a new situation. The flat-faced PAR38 lamp, the source in many downlights, is of course harder to grasp than the pear-shaped general-lighting-service bulb, particularly as the latter is likely to have some space between it and the reflector, whereas the reflector type can have a close-fitting housing.

Wherever special equipment is needed, or interruption in the normal use of the interior is unavoidable, there is added justification for group replacement of lamps. In any case, this practice is being adopted increasingly on other grounds. Discharge sources, including fluorescent tubes, are usually replaced in bulk at a calendar interval (often that corresponding to the nominal life), but tungsten-lamp replacement is better related to the failure of a given proportion, such as 20%. While this figure was regarded as economical, spot replacement of the lamps that had failed was necessary. Now that the tendency is to think in terms of a lower figure, it may be reasonable to accept the 'outages' until the bulk change is made—everything depends on the size of the installation and how conspicuous the failures are.

Two approaches to keeping maintenance costs down are to use lamps with long lives and to specify equipment that shows little depreciation in performance due to dirt. We can look at these in turn.

General lighting service (g.l.s.) tungsten lamps have a nominal life of 1000h. It can be increased by underrunning. Operating a 240V lamp on 220V roughly doubles the life, but it also means a loss in light output of about 15%. Where the circuit involves a dimmer and the setting is rarely at maximum, the life will be extended. Just occasionally running lamps in series can be worth while. Suppose a pair of outside wall lanterns are largely for decoration and little functional illumination is expected (the fitting might be a Scandinavian type with a smoked-glass globe). If long burning hours are anticipated, say, dusk to midnight, the orthodox choice of a 25 W clear bulb would need replacing about twice a year. But if the fittings are wired in series and each is given a 200 W clear lamp, the light output is about the same (with 120 V across each), and, barring accidents, the lamps should last as long as the fittings. The larger bulb, larger filament, and warmer light are probably all welcome. The main loss is in that each lamp is now consuming about 70 W. This kind of arrangement is theoretically uneconomical but, in practice, may seem justified. A better case might be made for running two 200 V lamps in series on a 240 V supply, or even three 110 V lamps.

Many types of tungsten–halogen lamp have lives of 2000 h, and some reflector lamps give 1500 or 2000 h. This is as a result of conscious choice in design (Chapter 11), and is a reflection of higher unit cost; so using tungsten–halogen instead of conventional tungsten (or reflector rather than g.l.s.) is rarely justified solely in terms of maintenance economics, unless the cost of access and labour considerably exceeds that of the replacement lamp.

'Blended' lamps have an internal tungsten filament in series with a mercury discharge source and so need no external control gear. The filament is underrun to give it a life comparable with that of the discharge tube, so the efficacy (lumens per watt) of the incandescent part of the system is low, but the relatively high red content is welcome as a supplement to the mercury spectrum. The total efficacy is only a little above that of the corresponding tungsten lamp, but the long life (6000 h in recent types) is often useful.

High-pressure mercury lamps and normal fluorescent tubes are usually regarded as having long lives; the nominal figure is 5000 or 7500 h. For cold-cathode tubing, on the other hand, it is 15 000 h—one of the main reasons for adopting this alternative lamp (in an inaccessible cornice, for example). The electroluminescent panel is another source with almost indefinite life, but one so low-powered that it is rarely used for general illumination; it can be valuable, however, for a continuously burning night light in a hospital or domestically, particularly as it will need almost no attention.

Sealed fittings, which prevent dirt from settling on the upper surfaces of the lamp and on the reflector, are likely to show good maintenance of their light output. Another approach is to encourage a strong flow of air through a fitting for its scavenging effect, either by using convection currents (as in high-bay industrial reflectors for 1000 W mercury lamps) or by giving the fitting an air-handling function; when air was first extracted through lighting fittings, fears were expressed that excessive soiling would result, but experience has shown that tubes and reflectors tend to keep cleaner than in conventional types.

It is natural to think that the main purpose of giving a lamp an internal reflector is one of light control. But an apparently incidental advantage is that the reflector, sealed within the envelope, is protected from dust, tarnishing, and mechanical displacement. If we mount a tungsten or mercury reflector lamp, cap up, in a dust-laden atmosphere, the front glass (pointing downward) stays relatively clean, and anything settling on the upper surface (the back of the reflector) has little effect on light output. The fitting need offer no more than support for the lampholder and any necessary protection from glare and impact. With fluorescent tubes too, dirty industrial conditions represent the major use of reflector types; these tubes have a diffuse coating over about two-thirds of the surface so that most of the light emerges from the remaining 'window'. Reflector tubes are available in only a limited range of ratings and colours.

The lumen method, the most widely used numerical technique in lighting design (it is described fully in Reference 1.2), depends essentially on the concept of utilisation factor as the ratio of the flux received on the working plane to the installed flux. In other words,

$$\text{flux received} = \text{flux installed} \times \text{utilisation factor}$$

This applies, as it stands, only in perfectly clean conditions. In reality, some correction is necessary for the effect of dirt. This is the 'maintenance factor', which is, in principle, the ratio of the average illumination actually recorded to the value that would have been achieved if everything remained scrupulously clean. The equation above then becomes

$$\text{flux received} = \text{flux installed} \times \text{utilisation factor} \times \text{maintenance factor}$$

To put it in another way, the product of utilisation factor and maintenance factor represents the ratio of what we get out of the system to what we put in.

The utilisation factor is affected by the design of the fitting, by the proportions of the room, and by the reflectances of its surfaces (we shall encounter examples in later chapters). Maintenance factor depends on the design of the fitting, the cleanliness of both location and activity, and critically, of course, on the effectiveness and frequency of maintenance. It is obviously meaningless to quote a maintenance factor without stating the cleaning interval for which it applies.

Yet schemes have often to be designed without any knowledge of the maintenance arrangements the building user will ultimately make. The time-honoured convention in the absence of any information is to assume of maintenance factor of 0·8. Suppose, as an example, that 600 lux is required in a room 8 m × 5 m. Since illumination in lux is the number of lumens arriving on each square metre,

$$\text{flux received} = 600 \times 8 \times 5 \text{ lumens}$$

From the equation discussed earlier,

$$\text{flux installed} = \frac{\text{flux received}}{\text{utilisation factor} \times \text{maintenance factor}}$$

The utilisation factor is discovered from the manufacturer's table giving its

Plate 16 Effect from outside of interior lighting may be important part of its function

Plate 17 Impression created by view into one space from another greatly depends on relative brightness

Plate 18
Conspicuous construction to house emphasis lighting in lecture theatre

Plate 19
Commercial 'spill ring' fittings recessed into domestic ceiling

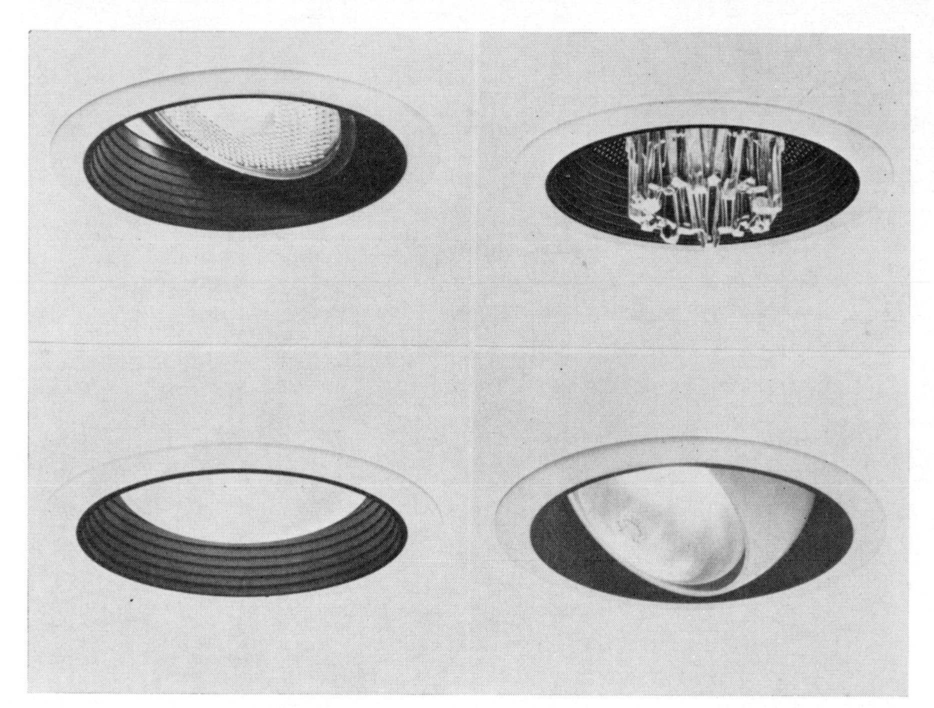

Plate 20 Recessed incandescent fittings with contrasting functions given a degree of visual unity by common elements

Plate 21 Related semirecessed and surface downlights

Plate 22 Thoughtful switch panel that simply but clearly distinguishes circuits by location and light source

Plate 23 Entrance hall of Commerzbank, Dusseldorf: ceiling designed around characteristics of bowl-mirrored reflector lamp

values for ranges of room proportion and surface reflectance. Suppose it comes to 0·5. Then

$$\text{flux installed} = \frac{600 \times 8 \times 5}{0{\cdot}5 \times 0{\cdot}8}$$

$$= 60000\,\text{lm}$$

We have then to work out an appropriate array of fittings with a total average lamp output of 60000lm.

If our conventional assumption of a maintenance factor of 0·8 is not justified, the resulting illumination will differ from the specified 600lux. If the figure that really applies is 0·9, the level will be about 670lux; if it should have been 0·7, 530lux. However, these are all averages, in both space and time. Even if the value of 0·8 is correct, and the average of 600lux achieved, the highest level (vertically under a fitting and immediately after cleaning and relamping) could be 900lux, and the lowest (between fittings, just before maintenance) might be half this, 450lux, or less.

Attempts have been made recently to put the selection and use of maintenance factors on a more sound and systematic basis (see, particularly, the IES Technical Report, Reference 2.9). Chapter 23, on industrial buildings, includes a further discussion. But the extent of the variation in illumination level, even in a fully and accurately calculated scheme, perhaps explains why uncertainty about maintenance factor has not worred designers more than it has.

Part 1 of this book has attempted to examine a number of lighting criteria so as to give some meaning to the adjectives in the generalisation advanced in Chapter 1, namely 'that, in most interior lighting, the aim is to produce an adequate illumination on relevant planes while limiting discomfort from glare, direct or indirect, to an acceptable extent and achieving a proper distribution of brightness generally; the light sources should have suitable colour rendering and the scheme should result in appropriate modelling; the degree of uniformity or diversity and the extent of simultaneous or sequential variety should be consistent with the situation; and the whole should be achieved at an acceptable cost in a way that permits effective and economical maintenance.' While discussing these criteria one by one has, I hope, contributed to an understanding of what is desirable in a lighting scheme, it must also have revealed how much they are interrelated.

There are no unique solutions.

Part 2

LIGHTING EQUIPMENT

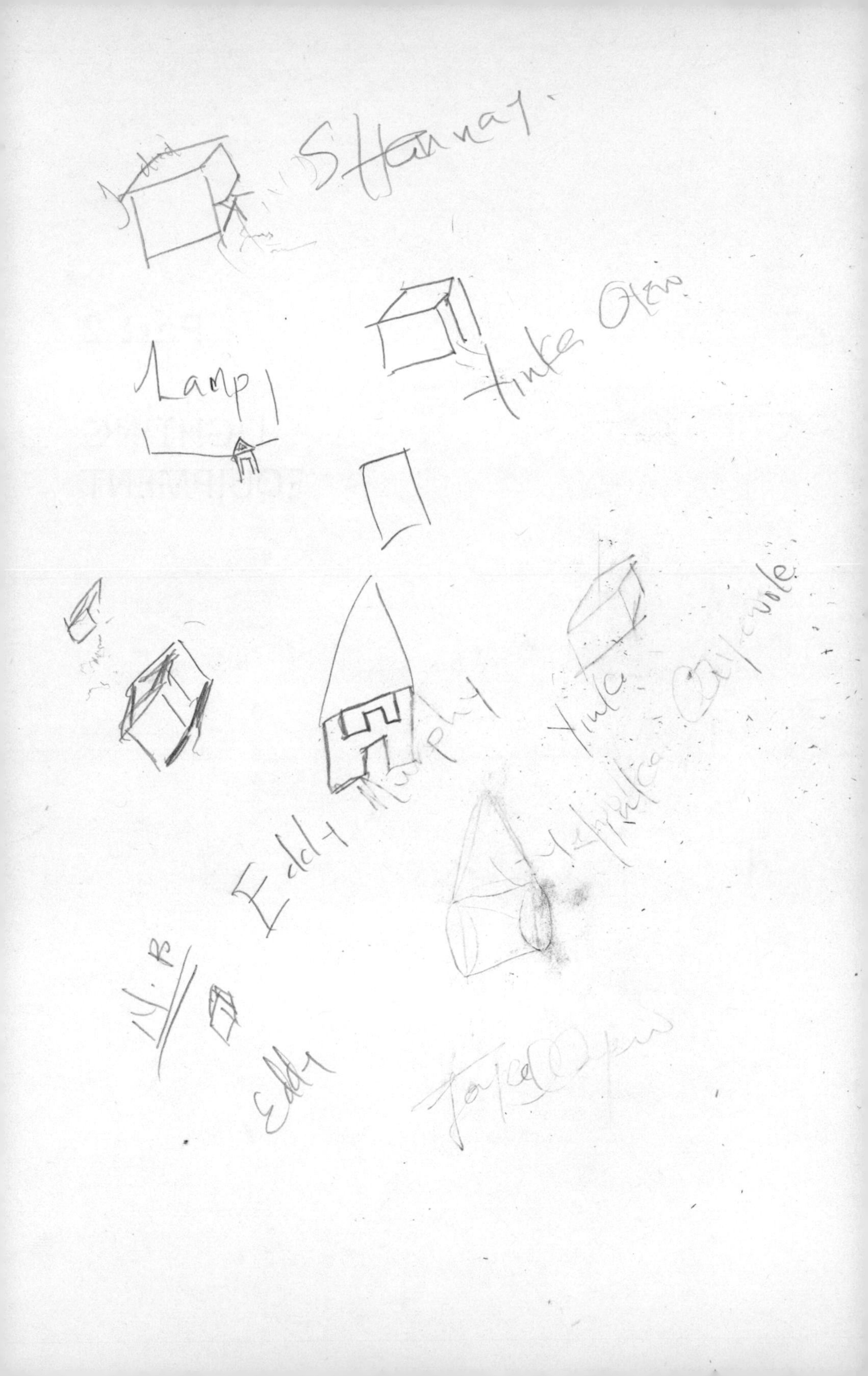

Lamp

Chapter 11

Lamps

Until recently, the choice of light source for an interior-lighting scheme lay between incandescent lamps and fluorescent tubes. I do not believe that the recent developments in discharge lamps have affected the situation to the extent that some of their advocates would claim. This is obviously my personal view, but it influences the treatment that follows to the extent that it consists mainly of a comparison of those characteristics of incandescent and fluorescent lamps that affect their selection; this is followed by a short discussion of high-pressure mercury lamps and other alternative sources. This chapter could not, of course, set out to be a comprehensive treatment of electric lamps. Most of what needs to be said about them emerges elsewhere in the book.

First, however, a brief account of the mechanism of light production. In the earlier type of lamp, the passage of electric current through a tungsten filament raises its temperature until it emits light. The hotter it is the more light it gives, and the whiter that light, but the evaporation of tungsten from the surface increases with temperature, so that life is shortened. The two main parameters in tungsten-lamp design are efficacy and life. Efficacy can be increased—i.e. more lumens per watt—if a shorter life is acceptable, or a longer life can be achieved with a fall in output. Both extremes are uneconomical, and the 1000 h life of the British Standard on general-lighting-service lamps (BS 161: 1968) represents the best value for money in typical circumstances. In early incandescent lamps, and in a few low wattage ratings now, the bulb encloses a vacuum around the filament to prevent it being oxidised, but most lamps of this type are 'gas-filled', the pressure of an inert gas within the bulb tending to inhibit evaporation from the filament. Convention currents flow within this gas filling and necessitate different filament configurations to avoid excessive cooling; in some types, the filament is coiled twice with this aim.

The main alternative mechanism of light production depends on fluorescence, the reaction shown by some substances, known as phosphors, that absorb ultraviolet (u.v.) radiation and re-emit energy within the visible spectrum. In a fluorescent lamp, the u.v. radiation from an electric discharge falls on a mixture of phosphors chosen to give the required output. Mercury vapour at low pressure is an effective medium for the discharge, and an extended tubular glass envelope is consistent with the optimum physical conditions. The inside of the tube is coated with the phosphor mix, usually of three components. The discharge itself produces a small amount of visible light in the separate lines of the mercury spectrum. The life to failure of a fluorescent tube is normally that of the electrodes, and there is no immediate relationship with lumen output as in tungsten lamps. Efficacy varies slightly with rating (watts) and loading (watts per metre), but is mainly influenced by colour-rendering quality: output falls as colour performance improves, so that the product of efficacy and colour-rendering index is roughly constant.

Table 1 Some characteristics of incandescent and fluorescent lamps

Characteristic	Incandescent lamps	Fluorescent tubes
Efficacy	10–15 lm/W	30–60 lm/W
Life	1000 h	5000 or 7500 h
Colour rendering	good, familiar	poor to excellent
Colour temperature	about 3000 K	choice: 3000–6500 K
Physical size	small source	relatively large
Heat dissipation	mainly radiation	mainly convection and conduction
Wattage range	25–1500 W	15–125 W
Control gear	none	necessary
Noise	none	slight
Flicker	none	slight
Temperature	little affected by ambient temperature	output temperature sensitive
Vibration	life reduced	little effect
Switching frequency	moderately frequent switching does not affect life	life reduced by frequent switching
Dimming	simple	more complicated

The comparative characteristics of incandescent lamps and fluorescent tubes have been summarised in Table 1. The following notes expand some of the points that appear there.

The 'lighting-design lumens' from a 100 W g.l.s. lamp (coiled coil, 240 V) number 1260 (i.e. 12·6 lm/W); the figure for a 1000 W lamp is 17300 (i.e. 17·3 lm/W). Fluorescent efficacies are around three times as great: the 40 W

Natural lamp gives about 2000lm (50lm/W, or 40lm/W if the 10W consumed in the control gear is taken into account) while the 85W, 2·4m White lamp yields some 6400lm from a total circuit wattage of 100 (i.e. 64lm/W). These high efficacies are the primary advantage of the fluorescent lamp and the main reason for its development.

There is a difference in the meaning of 'life' as applied to lamp types. For incandescent lamps, the term means life to failure, i.e. to the point at which the lamp ceases to give light; but, for discharge types, including fluorescent lamps, life to failure is so long (unless switching is unusually frequent) that the fall in light output means that the optimum use of the lamp has by then been exceeded. The quoted nominal life (7500h for most ratings in common use) is best regarded as a recommendation for an economical replacement period (the cost of putting in a new tube before the previous one has failed being justified by the extra light the new tube produces). Here too, fluorescent tubes have a definite edge on incandescent lamps (with relamping in a typical office after perhaps three years rather than five or six months).

Although the colour characteristics of incandescent lamps are widely acceptable, there is little we can do to modify them apart from overrunning or underrunning them or applying a filter. Fluorescent tubes, on the other hand, offer a wide range of colour types. We can regard the selection of one as the result of making two almost independent choices: of colour temperature and of colour-rendering quality, i.e. vertical and horizontal choices in terms of Table 2.

The small size of the filament in the incandescent lamp means that the light distribution from it may be precisely controlled. In particular, a concentrated

Table 2 Some fluorescent tube colours

	Colour temperature	Natural or familiar source in same colour range	High-efficacy	De luxe	Special
Approximate relative efficacy			100	75	50
CIE colour-rendering index			50–60	75–80	90–95
	6500K	overcast sky	—	Colour Matching/ Northlight	Artificial Daylight
	4000K	afternoon sun	Daylight	Natural	Trucolor 37 Kolor-rite
	3500K		White	—	—
	3000K	tungsten lamp	Warm White	De Luxe Warm White	Softone 27

beam may be produced, something impossible with extended tubular sources (this is not to imply that the optical design of fluorescent fittings may not be sophisticated).

The energy supplied to an electric lamp is converted into light and heat. Fluorescent lamps are sometimes regarded as cool sources; this can be a misleading idea, but it arises, first, because fewer watts are needed to produce a particular number of lumens, and secondly, because the thermal energy appears mainly as convected and conducted heat rather than as radiation. The precise proportions depend on lamp rating, colour, and so on, but typical figures for percentage energy dissipation are given in Table 3. The fluorescent figures

Table 3 Percentage energy dissipation from fluorescent and incandescent lamps

	Fluorescent	Incandescent
Convected and conducted heat	60	30
Radiant heat	25	60
Light	15	10

include heat from the control gear. Another note that should be added concerns the percentage of the input energy that appears as light. The 10% that is converted into light by the incandescent lamp has a spectrum with a pronounced bias at the red end, where the eye is relatively insensitive, whereas a typical fluorescent lamp has a spectrum peaking in the middle, in the yellows and greens, to which the eye responds strongly. Thus 15 W of 'fluorescent light' may be three times as many lumens as 10 W of 'incandescent light'. It may therefore be more useful to re-express the figures for the same flow of light in each case, as in Table 4. Once again, the range of efficacies means that these can

Table 4 Typical heat dissipation per 1000 lm

	Fluorescent	Incandescent
Convected and conducted heat	15 W	24 W
Radiant heat	6 W	48 W

be only representative figures; but they show that, in typical circumstances, we may have six times as much radiant heat from incandescent lamps as from fluorescent tubes with the same light output.

A final point on energy is that all the watts supplied ultimately become heat as the light is absorbed in room surfaces. Only to the extent that some light may escape through windows need we modify the statement that ten 100 W lamps represent the same heat input to a space as a 1 kW fire.

Let us continue to develop briefly the ideas summarised in Table 1. We have looked at incandescent and fluorescent lamps comparatively, in terms, so far, of efficacy, life, colour qualities, physical size and heat dissipation.

The extended wattage range of tungsten sources is often an advantage. Whereas the most powerful fluorescent tube in common use is 125 W (and 2·4 m long) a filament lamp can concentrate 1000 or 1500 W within a relatively small bulb. Incandescent lamps are connected directly to the mains supply; the control gear necessary to the operation of any discharge source adds inevitably to the complexity and cost of a fluorescent installation. Functioning correctly, a tungsten lamp produces no subjectively detectable sound or flicker. All inductive control gear produces some noise, though it is usually unnoticeable. Slim-sectioned batten fittings tend to emit a louder hum than those with more substantial ballasts, but a great deal depends on the mounting of fitting or gear—a piece of shopfitting or an area of suspended ceiling can act as a sounding board and acoustically amplify what would otherwise be negligible. Ambient noise is naturally relevant: a hum might be worrying in a library or hospital ward in the country that would be unnoticed in a city-centre office. Flicker is another minor characteristic of discharge sources that occasionally demands attention, though complaints about it are rare today. The need to consider it in particular building types is discussed later (see Index). It is exceptional for either noise or flicker to influence the choice of light source, since precautions against them are normally effective if fluorescent lamps are preferable on other grounds.

Conventional fluorescent lamps are very sensitive to temperature. Most types are designed for maximum output at an ambient temperature of 25°C; at 45°C, the output may have fallen by 15%, and it is similarly affected by low temperatures. The recently developed 'amalgam' tube may overcome some of the problems of heat buildup within enclosed fittings, but at temperatures such as those, say, in ovens, incandescent lamps will certainly be preferred. In exposed exterior positions, fluorescent fittings need features to help them retain the lamp heat (such as sealed enclosures or sleeves around the tube), and their use in situations such as cold stores is arguable—though high-loaded lamps such as the 1·5 m 120 W tube have a value here.

The advantage lies with discharge sources if the lamp is subject to vibration or excess voltage—these conditions represent the main causes of premature failure in tungsten lamps. On the other hand, since the life of a fluorescent lamp depends as much on the number of times it is switched as on aggregate burning hours, if a light is needed for minutes rather than hours at a time, an incandescent source may be preferable. Finally, simple voltage reduction is effective in dimming tungsten lamps, but more involved circuitry and specially selected tubes are required if a fluorescent scheme is to be dimmed (perhaps I have been unlucky in my experience of such installations, but I remain apprehensive about noise and flicker).

After this brief comparative survey of incandescent and fluorescent sources, let us turn to the high-pressure mercury lamp. A discharge in the low-pressure mercury vapour of a fluorescent tube results in energy being radiated largely in the ultraviolet region, with only a small quantity of visible light; but, in the small, high-pressure, quartz arc tube at the heart of a 'mercury lamp', the proportion of energy within the visible spectrum is very much higher, enough to represent an economical light source in itself. The radiation occurs, however, just at a number of specific wavelengths—the 'lines' of the mercury spectrum. This discontinuity, and an absence of significant lines at the red end, give the 'bare' or 'uncorrected' mercury lamp very poor colour rendering (the apologists for its early use in streetlighting used to say, 'It is better to look like a corpse than to be one'). If the inside of the outer glass bulb is coated with a phosphor that absorbs some of the ultraviolet from the discharge and emits red, a degree of colour correction is achieved. The 'fluorescent bulb', 'colour-improved' or 'colour-corrected' mercury lamp has for years been widely used in streetlighting and floodlighting, and to some extent in shop windows and industrial interiors. Recently, new phosphors have brought about a significant improvement in colour rendering, and the new lamps—with trade names such as Power White and Kolorlux—are being sold as appropriate for a much wider range of situations. Whether the improvement in colour performance is sufficient to justify this claim must be a personal conclusion based on experience of the resulting visual environment. Perhaps it would be fairer to say that the range of interiors that may happily be lit by mercury lamps has obviously grown; it is the extent of that increase that has still to be established.

Colour apart, the characteristics of mercury lamps can be outlined briefly, in the terms of Table 1. Output per watt lies broadly between those of de luxe and high-efficacy fluorescent tubes, and lives of the same order are achieved. Although the source is effectively the whole area of the outer, coated bulb (and so larger than a filament), it is still fairly small, and beam control by reflector is significant. Heat dissipation compares with fluorescent lamps, but a much wider spread of ratings is available: 50–2000 W. Control gear is needed; it is rarely noisy. Flicker is perhaps slightly more noticeable than with fluorescent tubes. Since the lamp operates with a fairly high temperature in its discharge tube, it is much less sensitive to changes in the ambient thermal conditions. For the remaining three characteristics in Table 1 (vibration, switching frequency, dimming), the entries there for fluorescent tubes may be repeated: 'little effect'; 'life reduced by frequent switching'; and 'more complicated'. However, a note should be added on running up: a mercury lamp takes several minutes after switching to approach full output (typically, 80% lumens after five minutes); further, if the electricity supply is interrupted, the lamp will not strike again until it has cooled down considerably (this can entail the obligation to include some subsidiary alternative sources in a mercury scheme).

Mercury-halide lamps, where the colour improvement is due to additives to the vapour in the discharge tube, are not much used indoors. The older 'blended' lamp, however, still has a usefulness. Since the current is limited by an integral tungsten filament in series with the discharge tube, no external control gear is needed. Recent versions have a 6000 h life, and some have a phospor coating. Efficacy is not much higher than for incandescent lamps, but the extended lamp-replacement period can be valuable where fittings are inaccessible. The high-pressure sodium lamp, with its golden-white light, has occasional application within buildings. Examples will be mentioned (see Index, again), and the many varieties and special types of both incandescent and fluorescent sources will be discussed as we encounter their use in particular building types.

The most useful background material on light sources for the designer is to be found in a series of booklets published by the Lighting Industry Federation (References 2.23, 2.24, 2.25 and 2.26). Manufacturers' literature covers much essential information. Unfortunately, a great deal of what is published on lamps makes very technical reading, and the usual inclusion of a short paragraph at the end on 'applications' suggests distorted priorities. But it is clear that lamps are the raw materials of lighting, and of great concern to anyone involved in its design.

Chapter 12

Fittings

A lighting fitting allows the lamp to operate, and controls the light that it produces. These two main functions may not be entirely separate; an enclosure that keeps moisture away from the electrical contacts may also act as a diffuser. But we can consider them almost independently.

The purpose of a lighting fitting is primarily optical. It exists to control the brightness of the source and the way the light is distributed. The other problems of fittings design—mechanical, electrical, thermal—must be resolved satisfactorily, but, in a sense, they are incidental. The central fact about a lighting fitting is what it does with the light produced by the lamp that it houses.

This optical performance can be expressed as the answers to a number of questions. Does the fitting allow a good proportion of the light from the source (the 'lamp lumens') to emerge? How does the amount of light going upwards compare with the downward flow? Is that downward flow widely dispersed or concentrated? What apparent luminous area does the fitting present to the onlooker?

The term 'light-output ratio' (l.o.r.) is used for the ratio of fittings lumens to lamp lumens. In a sense, this expresses efficiency, but the term is not a good one to use, since a fitting with a high l.o.r. is not necessarily going to do a particular job as efficiently as another with a lower l.o.r. but a more appropriate distribution.

Suppose a fitting takes a lamp with an output of 5000 lm. Photometric measurements of the intensity at regular angular intervals allow the polar curve to be plotted; from this, the number of lumens in each 10° zone around the fitting can be calculated. The result might be

total in upper hemsiphere	1500 lm
in lower hemisphere	2000 lm

We can express this either as a total light output ratio or separately for upward and downward components:

u.l.o.r. = 0·3
d.l.o.r. = 0·4

Notice that the upward light-output ratio is that of upward light from the fitting to the light from the *lamp*. The 'upper flux fraction', on the other hand, is the ratio of upward light to the total from the fitting. Thus here

$$\text{upper flux fraction} = \frac{1500}{3500} = 0{\cdot}43$$

$$\text{lower flux fraction} = \frac{2000}{3500} = 0{\cdot}57$$

The sum of the flux fractions is unity, but the sum of u.l.o.r. and d.l.o.r. is the total light-output ratio.

For many purposes, it is the ratio of upward to downward flux that is important, and this is expressed as the 'flux-fraction ratio' (f.f.r.):

$$\text{f.f.r.} = \frac{\text{u.f.f.}}{\text{l.f.f.}}$$

In the example that was taken,

$$\text{f.f.r.} = \frac{1500/3500}{2000/3500} = \frac{1500}{2000} = 0{\cdot}75$$

This is clearly the same as

$$\frac{1500/5000}{2000/5000} = \frac{1500}{2000} = 0{\cdot}75$$

That is, by definition,

$$\text{f.f.r.} = \frac{\text{u.f.f.}}{\text{l.f.f.}}$$

but, by deduction,

$$\text{f.f.r.} = \frac{\text{u.l.o.r.}}{\text{d.l.o.r.}}$$

All this means that it is only necessary to state the upward and downward light-output ratios to give, by implication, the total light-output ratio, the flux fractions, and the flux-fraction ratio—i.e. to answer the questions about the proportion of the lamp lumens that emerges and how much goes up and how much down.

Suppose, for instance, the u.l.o.r. is 0·50 and the d.l.o.r. is 0·25. Then the l.o.r. is 0·75 and the flux-fraction ratio 2.

One useful, and long-standing, classification of fittings depends directly on the upper flux fraction, though it could also be expressed in terms of flux-fraction ratio (Table 5).

The other important type of classification depending on optical performance relates to the spread of the downward light from the fitting; i.e. it provides an answer to the question as to whether the downward flow is widely dispersed or concentrated. The British Zonal System is fully described in IES Technical

Table 5 Classification of lighting fittings according to proportion of upward and downward light

U.F.F.	Class of fitting	F.F.R.
0 –0·1	direct	0 –0·11
0·1–0·4	semidirect	0·11–0·67
0·4–0·6	general diffusing	0·67–1·5
0·6–0·9	semi-indirect	1·5 –9
0·9–1	indirect	9 –∞

Report 2 (Reference 2.2) and briefly in 'Interior lighting design' (Reference 1.2). The system makes it possible to assign most fittings to one of ten categories, i.e. to give a fitting a 'BZ number' from 1 to 10, the number rising as the spread of downward light becomes more dispersed. A spotlight with a vertical beam would be BZ1, and a freely suspended vertical fluorescent tube BZ10 (we shall meet other examples). Strictly, the BZ number is a function not just of the fitting but also of the space it lights, and, for some fittings, the number may change at a particular value of the room index expressing the proportions of the space. Thus the notation BZ3/1·5/BZ4 means that the fitting to which it applies acts as BZ3 in small rooms, but, for spaces with a room index of 1·5 or more, as BZ4. Knowledge of the BZ number may be needed to determine the limiting spacing/mounting-height ratio (Chapter 7); it is also essential in the calculation of the glare index of a proposed scheme.

Another essential piece of data in finding a glare index is the luminous area. This is the apparent area of the luminous part of the fitting as seen from directly beneath it, except for types classified as BZ9 or BZ10, where the very wide dispersion means that the view from the side is more important. Since quoting luminous area in square metres (m^2) or square millimetres (mm^2) can result in awkwardly small or large numbers, it is usually given in square centimetres (cm^2).

The data that we may reasonably expect fittings manufacturers to quote in their literature include

(i) u.l.o.r. and d.l.o.r.
(ii) BZ number
(iii) luminous area
(iv) limiting spacing/mounting-height ratio for conventional uniformity
(v) table of utilisation factors for various values of room index and ceiling and wall reflectance.

Whereas the fittings maker needs to plot the polar curve to deduce some of these data, there is little point in reproducing it for the user, except as a sort of technological garnish.

To illustrate some of these idea, let us compare a batten fitting for a single tube with the result of adding three different attachments (Fig. 9). From left to

Table 6 Some fittings data

	a	*b*	*c*	*d*
u.l.o.r.	0·23	0	0·15	0·20
d.l.o.r.	0·70	0·80	0·65	0·50
BZ number	7	4	5	6
Luminous area, cm²	1440	3120	3120	2640

right, *a* is the basic spine, lampholders, and tube; *b* has an unslotted, closed-ended metal-trough reflector added; in *c*, the attachment is an open-ended trough of similar section but in plastics, transmitting some light; and *d* is an enclosed opal diffuser.

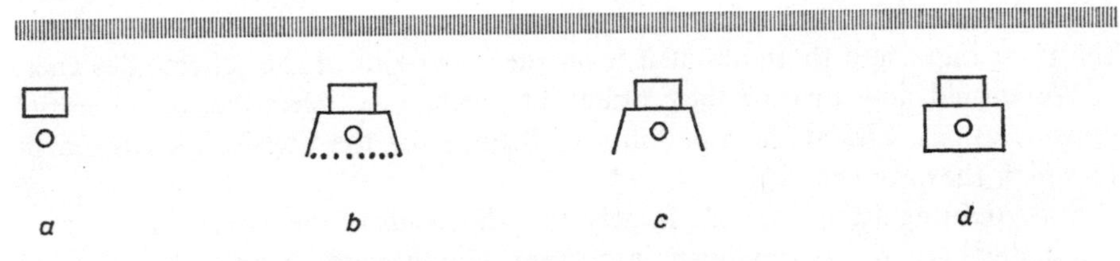

Fig. 2.1 Batten fittings and attachments

It will be noted from Table 6 that the total light ouput ratio is highest (93%) for the fitting with the least light control (the bare-tube batten *a*) and least (70%) for the enclosed diffuser *d*. There is most downward light for the metal reflector *b*, and also the narrowest downward spread BZ4.

The flux fraction ratios may be calculated: *a*, 0·33; *b*, 0; *c*, 0·23; and *d*, 0·40. This confirms that the metal reflector is a direct fitting, and the other three all semidirect.

There are, of course, many other comparisons we might make, but, before leaving this group of fittings, let us look at their utilisation factors in a room of medium proportions (room index = 2·5) with a light ceiling and a fairly dark one.

Table 7 Some utilisation factors

	Light ceiling (ρ = 0·75)	Dark ceiling (ρ = 0·30)
a	0·62	0·51
b	0·60	0·58
c	0·57	0·52
d	0·52	0·40

It appears that, though the metal reflector *b* gives most downward light, the batten *a* has a higher utilisation factor, provided that there is a light ceiling to throw back some of the upward light. Conversely, the effectiveness of the metal fitting is least influenced by the change to a dark ceiling, though the fact that there is a change in the utilisation factor (from 0·60 to 0·58) shows that some part of the effect in the first case is due to multiple interreflection. As we should expect, it is the fitting with the highest flux-fraction ratio, the enclosed diffuser, that is most affected by the darker ceiling.

My concern, in this book, is with the application of lighting fittings rather than with their design. The whole subject is examined fully in Reference 2.27; I have attempted in the journal of the Royal Institution of British Architects [1972, **80**, (5), pp. 202–204] a systematic examination of the selection of lighting fittings in terms of the CIB's 'Master list of properties for building materials and products'. The use of fittings is, of course, a constantly recurring topic throughout these pages. In this chapter, we have so far looked mainly at some of the language for expressing the optical performance of a fitting, since this is the most important thing about it from the viewpoint of the scheme designer.

We should now turn to look briefly at mechanical, electrical and thermal considerations, and at the suitability of fittings for the physical environment in which they are installed.

Since lighting fittings operate largely over the heads of the users of a building, mechanical security is obviously important. Components have to be removed for maintenance, and a really positive indication that they have been replaced correctly is welcome.

The main reference, in the UK, is BS3820 : 1964, 'Electric lighting fittings'; similar standards exist in other countries. Section two of the specification covers constructional and dimensional requirements, but precise tests in this area are difficult to formulate, and the text reads in places more like a pious hope than a specific requirement: 'Precautions shall be taken to prevent the lamp and any part of the fitting being dislodged by vibration or other adverse conditions which may occur in service .. compliance is checked by inspection...' However, the difficulty is mainly one of dealing with the general case. Provided that one bears the need in mind, it is usually possible to satisfy oneself that confidence in the mechanical security of a particular fitting is justified.

Electrical and thermal requirements are established in Sections 3 and 4 of BS3820, and here it is possible to set down much more precisely the conditions that should be met. In the ordinary way, stated compliance with BS3820 should leave the specifier of lighting fittings completely confident in the electrical and thermal characteristics of the equipment—this, after all, is the function of such standards.

However, it is as well to be aware of the major hazards of inadequate thermal and electrical design, if only to avoid dangerous situations when lamps are integrated with builder's work, or when a 'one off' special fitting, which will never go near a test house, seems the only answer to a particular lighting problem. Here the requirements of standards such as BS3820 form a sort of check list: cable entries, cord grips, adjusting devices ('.. so constructed that cords or cables are not pressed, clamped, damaged or twisted ..'), wireways, internal wiring ('.. suitable type and size ..'), terminals, earthing, live parts,

Plate 24 Lighting equipment suspended completely free of structure

Plate 25 Where building reveals its basic structure, physical integration of lighting hardware is difficult; the approach that applies fittings explicitly but logically may be more successful. Here temporary exhibition is lit by spotlights fitted to trunking and track mounted beneath structure

26

27

Plates 26 & 27 Making lighting equipment part of room means that changes in effect seem less arbitrary

Plate 28 Directional lighting from incandescent sources adds emphasis and reveals modelling

Plates 29 & 30
Troughs for fluorescent lamps follow line of building at Sunderland Civic Centre

29

30

Plates 31 & 32
Recessed downlights in two forms of suspended ceiling; the one with fissured tiles and without strong directional effect accepts circular apertures easily; the other is boarded and benefits from square mountings provided

31

32

and so on. Thermal and electrical hazards are, of course, closely interrelated since all light sources develop heat; inadequate dissipation leads to excessive temperature rise, the breakdown of insulation, and so—full circle—to a fire risk.

Fittings may be classified in many ways other than in terms of optical performace. Very often, we speak of one as a fluorescent or an incandescent fitting, referring of course to the light source it houses; or the building type for which it was designed may help to identify its characteristics—a hospital fitting, domestic types. Materials too are often invoked—an 'all-plastics' fitting, 'lighting glassware'. How the fitting is mounted is another important feature—recessed, semirecessed, surface, pendant, portable.

A very important classification depends on the conditions in which the fitting is intended to be operated. An obvious distinction is that between interior and exterior types, but even here, there are finer shades in 'under-canopy' fittings and so on. BS3820 recognises eight classes according to the degree of protection against moisture or dust (ordinary, drip-proof, rainproof, jetproof, watertight, submersible, dustproof, dust-tight), and three according to the degree of protection against corrosion (indoors, normal atmospheres; outdoors, or indoors in atmospheres of high humidity; and chemically corrosive atmospheres). Fittings may also be classified by the type of protection against electric shock ('all-insulated', 'double-insulated' and so on) and in terms of their suitability for operation in atmospheres representing some degree of explosion hazard (there is a separate standard, BS889 : 1965, for flameproof lighting fittings).

In their extent and elaboration, these many systems of classification may seem overdeveloped when summarised in general terms in this way. But we shall see (in Part 4) examples of their value. It is, of course, useful to be able to specify simply 'dustproof to BS3820', for instance, and to know that this has a precise meaning.

Chapter 13

Other lighting equipment

Cable trunking is more a piece of electricity-distribution equipment than a lighting device, but it has an immediate application in many lighting schemes. Particularly where rows of fluorescent fittings are to be suspended beneath an open-structure roof, the runs of trunking provide the equivalent of a ceiling plane, they ensure inherently good alignment of fittings, and they facilitate a rearrangement of the scheme or additions to it. The trunking may, of course, carry other services. Even with a flat ceiling, there are advantages in reducing the number of fixings to the structure and in ready access to the wiring from beneath. Recessed flanged trunking can be used to support ceiling panels or tiles.

Lighting track can be regarded as a development of trunking, but usually the conductors are a rigid and integral part of the unit lengths. The track provides complete flexibility of fittings positioning in one dimension, and parallel runs give considerable freedom of choice over a plane. The various proprietary tracks claim particular advantages, and the range of effects possible by simple switching has been increased by the introduction of 2-circuit and even 4-circuit tracks. However, it is worth pointing out one essential difference that remains between display lighting and stage lighting: the former uses a small group of circuits each with numbers of fittings, and the latter has many circuits with only one or two lanterns on each; so separate outlets, rather than track, remain typical of stage installations.

Switches of modest current capacity are used so extensively in domestic wiring that bulk production makes them cheap; so it need cost no more to split up a 4kW tungsten load into four circuits each with a 5A switch than it does to provide a single control for the lot—and the advantage in flexibility is obvious. Remember that ratings are for a resistive load, such as tungsten lamps, and that fluorescent circuits need switches rated at about twice the nominal load of lamps plus gear (the manufacturer usually quotes precise limits).

Switching devices depending on relays have the primary advantage that the main current through the appliance, here the lamps, does not flow through the equipment at the switching position, so that a large load can be controlled by switching a small one. But relay techniques have been developed to give greatly increased subtlety of control, particularly in offices and hotels (Chapter 28).

The spread of dimming into the home is significant of the growing demand for a wider range of lighting effect. It has led to the marketing of many small dimmers at remarkably low prices, but it is only realistic to accept that there are major differences in quality in domestic and professional ranges; some domestic types seem to have short lives. If the budget restricts one to the modest level, two precautions are worth taking. First, use a dimmer with apparently spare capacity (such as one nominally for 500 W for a 300 W circuit), and, secondly, do not provide a separate accessible switch, so that the operator uses the dimmer control itself as the switch and thus the circuit is always brought in progressively from zero, thereby avoiding the heavy initial surge of current that overloads the unprotected equipment and causes failure.

Although it is hardly lighting equipment, this is probably the place to mention alternatives to the 50 Hz, 240 V supply normal in buildings in the UK. Low-voltage (or more correctly 'extra low voltage') supplies at 24 V or 12 V imply heavy currents, and so are usually localised. Where display equipment uses 12 V lamps, each fitting, or small group of fittings, tends to have its own transformer—indeed, many modern spotlights have integral transformers. The exception appears in a 12 V d.c. supply for emergency lighting coming from a battery room. Incandescent lamps are available for this voltage, but fluorescent lamps need special control gear (a 'transistor ballast'), which, in fact, produces alternating current at a higher voltage. Similar equipment is used in small boats and caravans, where again the supply is d.c. from a 12 V battery, but this is really outside the scope of a book on building interiors. The low-voltage emergency supply running right round the building and so demanding some duplication in wiring is less common than it was, as emergency lighting today depends increasingly on fittings that normally operate conventionally but contain an internal battery that takes over when the mains supply is interrupted. Local-authority demands for emergency lighting vary, so that the particular requirements need to be established.

If fluorescent lamps are operated on a high-frequency supply (say, 8000 or 10 000 Hz rather than 50 Hz), the control gear can depend on capacitive rather than inductive devices, with considerable savings in weight and some in cost. The efficiency is higher by around 10 %, flicker disappears, and noise problems are likely to be less (depending on the precise frequency). At the moment, there are problems in producing the high frequency economically from the normal supply, but it seems certain that these advantages will be exploited in large buildings, even if it means an extra service to be distributed (electricity for fluorescent lighting in addition to the 50 Hz supply for other purposes). An alternative that avoids this last difficulty is the production of the high frequency within each fitting by a device that becomes part of its control gear; progress in electronics could make this economical.

In these three chapters on lighting equipment, which make up the short second part of the book, I have attempted to comment on some of the main points to be borne in mind in applying lamps and fittings to the job of lighting building interiors. Specialist works, such as References 2.23–2.29, must be consulted for more comprehensive information, and the Index may be used as a guide to related topics in other parts of the present volume.

Part 3

ELECTRIC LIGHTING IN BUILDING DESIGN

Chapter 14

Daylighting

The headings of the chapters that follow—'Form', 'Fabric', 'Sound' and 'Heat'—do not imply any general discussion of these aspects, but rather a look at the inescapable way they are linked with lighting, so that decisions on lighting design must not be taken without an awareness of the consequences on the shape of the building, on its structure, on acoustics, on the thermal design—and vice versa. For the purposes of discussion, there is some attempt to consider these relationships one by one, but we shall find that this is almost impossible. The form of a building and its thermal characteristics, for example, are not only both related to its lighting, but clearly to each other as well.

This chapter, in the same way, does not treat of daylighting in itself (the subject is, of course, central to architectural studies, and has an extensive literature—see, for instance, References 1.1 and 2.30–2.33). What we are concerned with here is the relationship between daylighting and electric lighting; this arises not only from their simultaneous presence for much of the time in modern buildings but also, more fundamentally, because our expectations of lighting generally are still largely conditioned by our experience of daylighting.

The traditional view of building lighting is that, while daylight is available outside, we arrange for enough of it to penetrate to the interior to permit the space to be used; only when daylight fades is artificial lighting employed, and then as a substitute for the real thing. For most of the history of building, this approach has applied, and quite justifiably in view of the inadequacy of available artificial sources. Some consequences of this situation, however, have become so deeply embedded in traditional thinking about building that we may need to remind ourselves that they are not eternal verities.

There are no cosmic laws to say that buildings must have windows, that they can be no more than two or three rooms thick, that a daylit building must have natural ventilation, or that seeing out is always desirable. Very often these ideas will be valid, but we can no longer assume them; the questions must always be put.

After nightfall, any building will need artificial lighting if it is to continue to function; but, during the day, three approaches are possible. The first produces fully daylit spaces, as in conventional domestic accommodation. In the second, the lighting for the activity housed is provided by a designed combination of daylight and electric light. And the third possibility is to depend on electric lighting completely for the working illumination, without preconceptions about whether daylight enters the interior for other purposes or about the extent of visual access to the exterior (this third option includes the windowless building, but embraces a much wider range of situations).

The first approach was assumed in most building before the middle of this century, and in some construction today—conventional primary schools, housing, and so on. We should not lightly disregard the weight of tradition that supports it. Although it is true that, in most developed countries, more energy is used for daytime lighting than for night-time lighting—so that the major use of electric light is while daylight is present—many people seem to feel spontaneously that to have the lights on all day represents either an unconventional solution or incompetent daylighting design.

The takeover of lighting during daylight hours in industry and commerce by electric lamps occurred before it had received much theoretical consideration. The reason lies in the high efficacy (output per watt) of the fluorescent tube. While the only source available was the tungsten lamp, the conventional lighting loading for an office block of 10–20 W/m^2 produced perhaps 50–100 lux—a not very significant addition to the relatively generous daylighting in a typically shallow, high-ceilinged, prewar building; but, when, around 1950, 15–20 W/m^2 of fluorescent lighting yielded 200–300 lux, this was a worthwhile supplement to daylight in the slightly deeper, rather lower office interior of that time. So people began to switch on the lights when they arrived in the morning and to leave them on all day.

In the middle 1950s, workers at the UK Building Research Station and elsewhere began to question the logic of designing schemes for night-time conditions when their major use was during the day. If a system were to be designed as permanent supplementary lighting—as a supplement, that is, to a building's daylighting—what, they asked, would it be like? The work of Hopkinson and Longmore at BRS was reported in a celebrated paper on p.s.a.l.i., 'The permanent supplementary artificial lighting of interiors', at a London meeting of the IES in 1959; it is reprinted in the second part of Reference 1.1, 'Architectural physics: lighting'. The summary may be quoted:

> 'Since the opening of the half-century, new ideas on building have been visible everywhere. These visible changes are the expression of a revolution in building construction methods and in the economic picture of building. Rising costs demand a more efficient use to be made of available space, which, in turn, means deeper rooms with lower ceilings. Daylight is not free, it must be paid for in terms of cubic feet of space, square feet of glass, and loss of heat. In fact, the provision of adequate daylight to the back of a deep room in the quantity now considered adequate by modern standards would lead to such an amount of glazing that not only would

the cost be prohibitive, but the discomfort from sky glare would be unacceptable.

'It is shown that permanent supplementary lighting can be used to overcome these difficulties. The cost of installing, running and maintaining such lighting is offset by the reduced costs of building and maintenance and by a more efficient use of the available site.

'Experimental studies have been made to determine the levels of supplementary lighting that are necessary for a good integration of daylight and artificial light. The results of these studies are applied to a new technical college and to an agricultural research laboratory.'

A model typical of those used in this work gave a reasonably detailed representation—at a 1/12th scale—of a teaching laboratory, some 8 m deep and with a 3 m ceiling height (Fig. 3.1). With windows along one side, this means a daylight factor in the area farthest from the window wall of about 1 %—inadequate by current standards. Supplementary light was provided in this area by two short tubes above white louvers flush with the ceiling; the local illumination produced was controlled by dimmers over the range 1–1400 lux. Observers were asked to choose a dimmer setting to give the best balance of brightness over the room; it is significant that this, rather than any question of adequacy of illumination, was the criterion. There was naturally a wide spread of results from one observer to another and with different sky conditions outside. The stress on the brightness balance explains what is sometimes still regarded as a paradox, that higher settings were usually chosen for higher sky brightnesses—in other words, the brighter the day, the more supplementary light is felt to be desirable.

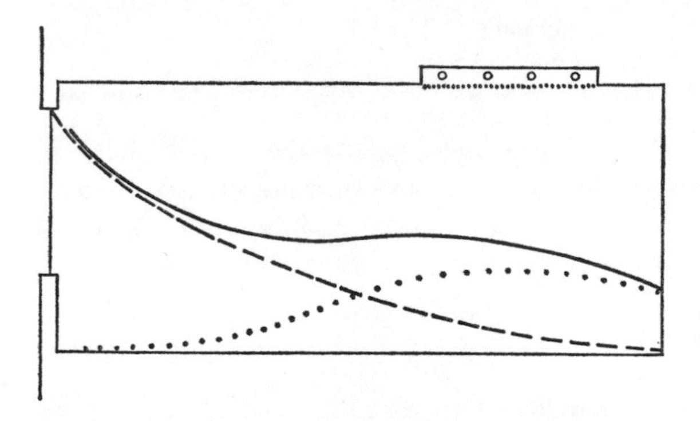

Fig. 3.1 Permanent supplementary artificial lighting of interiors (p.s.a.l.i.)

In spite of the spread shown by figures for what was considered the optimum setting, most people under most sky conditions were reasonably satisfied with supplementary levels in the 400–600 lux range. At the time, this was about double the level of afterdark illumination associated with good practice in offices and similar interiors. In a number of applications of these ideas, the switching was arranged to result in the same electric load by day and by night. A scheme covering the whole space uniformly might produce 250lux after dark; during daylight hours the fittings in the half of the room nearer the windows would be out of circuit, but double the number of tubes would be switched on in the other half to give a local supplement of about 500lux (Fig. 3.2).

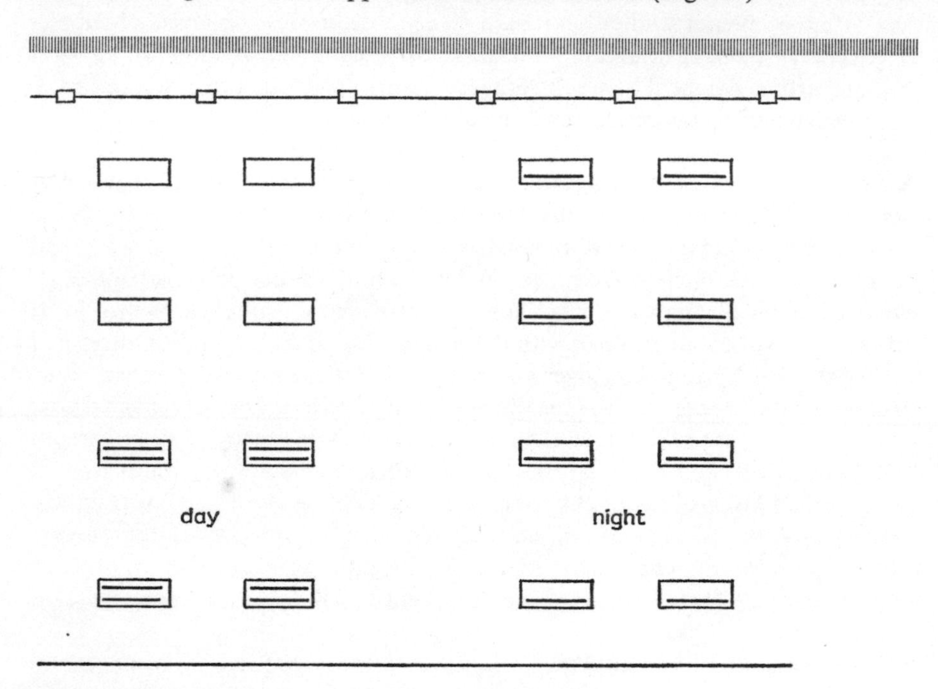

Fig. 3.2 Switching in typical p.s.a.l.i. scheme

A number of variations are, of course, possible. At the Esso building in Victoria Street, London, for instance, the three rows of fittings (A, B, C—with A nearest the window wall) have two or three tubes each, with the switching again arranged to give the same loading per bay by day and by night:

day	A: (none)	B: 3 tubes	C: 3
night	A: 2	B: 2	C: 2

However, the continuing rise of general or night-time illumination levels meant that, in about a decade—by the late 1960s—500 or 600lux was quite common in overall schemes, so that simple switching by rows parallel to the window wall was enough to permit the application of p.s.a.l.i. ideas in rooms of medium depth (say 7–10m). Further rises tended to come at the same time as the adoption of large open-landscaped offices of considerable depth; 700–1000lux overall by day and by night is now not uncommon. P.S.A.L.I. as such was no longer relevant, as the electric lighting in this situation could not be considered supplementary. Thus it was that we began to talk of p.a.l., or permanent artificial lighting, and it was suggested that p.s.a.l.i. was merely a passing phase in building lighting, between daylighting in traditional structures and full electric lighting in progressive designs. But this was oversimplification. P.S.A.L.I. may not have proved to be such an important element in office lighting as once seemed likely, but it retains its relevance in rooms of intermediate depth, and its basic tenet of a balance between daylight and electric light designed to produce the impression of a predominantly daylit interior can be useful in a wide range of building types from housing to hospitals.

A clear-cut opposition between daylighting and electric lighting appears in the single-storey factory, which may have an array of roof lights or an opaque roof with an electric system for permanent use. It was during the 1950s again that interested parties, such as the Electricity Council in the UK, began to make what still seemed the surprising suggestion that a building using artificial lighting throughout the working period could be more economical than one depending on 'free' daylight.

In any comparative costing, we need to consider initial costs as well as running costs, and it is necessary to make assumptions about the life of the building and of the lighting installation, to reduce first costs to an equivalent annual sum. The items to be considered for each of the alternatives include structure, maintenance, lighting installation, electricity charges for lighting, lamp replacement, heating installation, and fuel charges for heating.

A simple shed-type roof with some translucent panels might be cheaper than a flat, opaque, insulated roof, but more sophisticated forms of daylighting giving better visual conditions inside are likely to be more expensive. Certainly, maintenance will cost more for any form of rooflit space.

The lighting installation need not differ much in cost in the two cases, for any factory will be in use for a considerable time when daylight alone is inadequate. The life of the installations is likely to be similar, since this is much more a question of the number of years the scheme has been use in than of the total time the lamps have been on. Obsolescence rather than disintegration is the usual stimulus for replacement. On the other hand, managements tend to specify a higher quality of lighting for a no-daylight space than they do elsewhere, perhaps out of a feeling that some compensation is called for.

The heating installation needs a much greater capacity when it has to cope with the considerable heat loss through a glazed roof, and fuel charges will be correspondingly greater.

When the lighting is in use all the time work is going on, electricity charges and lamp replacement costs rise, but in less than direct proportion to the hours of use; this is because part of the electricity bill is a standing charge, usually depending on installed load, and also because less frequent switching and foreknowledge of the utilisation of the lighting permits greater precision in planning group lamp replacement and other maintenance operations.

Therefore the amounts by which the costs of structure, maintenance and heating for the daylit factory exceed those for the alternative represents the cost of daylighting, to be set against the additional expense of running the lighting for longer periods. A study carried out by the Pilkington Research Unit at Liverpool University (Reference 2.34) showed, in 1962, that the balance was delicate and that no sweeping generalisation was possible. The conclusion from any given set of assumptions depends very much on the standards of daylighting and electric lighting adopted, and on electricity tariffs (this last factor could be one reason for the much wider adoption of the opaque-roof factory in the USA, where energy is generally cheaper than in Europe). Now a decade later, one's feeling is that there still needs to be some special justification, in the product or its processing, for a factory with no daylight, but that, paradoxically, most factories will use electric light continuously. This is, in a sense, permanent supplementary daylighting, where there is the best of reasons for the supplement, namely that people like having it even though it costs money.

Although the question of no-daylight interiors is of increasing importance, it is better for us to consider it mainly in the context of particular building types, in Part 4. What can usefully be discussed briefly and in general terms here is its obverse, i.e. the functions of a window, and also the expectations that we have about lighting during the hours of daylight. Some study of natural lighting is our best guide to design for its absence. The problems of windowless environments are examined in a Greater London Council (GLC) research paper with that title (Reference 2.35), which is perhaps of greatest value as a guide to further sources.

In this century, windows have enjoyed two liberations. Early in the modern movement, they were released from the structural constraints that applied while they remained apertures in load-bearing walls. More recently, they have been freed from the obligation of providing working illumination. The effect of this has been to prompt a reappraisal of the window's function, almost as though its role were being considered for the first time. The child's idea that a window is primarily for looking out of is not far from the truth. Certainly, a major function is the transmission of visual information. Conventionally, this implies presenting those within the building with a view of the exterior scene. That view is of course affected by the size and shape of the window, and there has been much discussion on the relative merits of horizontal strips, vertical bands, square openings, and so on. Once again, any generalisation is unconvincing, since the best shape must be related to the characteristics of the external scene. We should recognise too that, from any position farther than a few metres from the perimeter, the view from a upper floor is likely to be limited to the distant landscape and the sky. However, even from rooflights, as long as the glazing is clear, we can tell what sort of day it is and whether to take a raincoat with us at lunchtime—the transmission of visual information.

A related, but separate, function of the window is to be found in the opportunity it provides for people to focus on a distant object. This is a physiological relief, for the ciliary muscles which control the curvature of the lens in the eye are relaxed when it is focused at infinity. There is, of course, psychological release too, in being able to see out from the enclosure housing the activity to the neutrality beyond, but this is a function of visual access to the exterior, i.e. view rather than distance. Inside a large building, such as a factory with extensive unobstructed production areas, the distant focus may be available within the space. It is common experience that windowless interiors are more acceptable where the scale is large. In some factories housing processes demanding closely controlled physical conditions, the offices at the perimeter have 'windows' giving views inward to the production space rather than outward to the open air.

It may be that what people most object to in a completely artificial environment is the unrelenting sameness of the visual conditions. Windows bring variety in two ways: in the changes that can be observed in the external scene, including the sky itself, with weather, season, and time of day; and in the effect of these changes on the appearance of people and things within the space or of the surfaces that define it.

Even at a distance from a side window at which its contribution to illumination on a horizontal working plane is almost negligible, there is still a significant flow of daylight in a direction close to the horizontal. The total flow (Reference 2.21) can be regarded as the resultant of two vectors, one vertical, from the array of electric-lighting fittings, the other nearly horizontal, from the window. The resultant will thus lie in some intermediate direction and in most circumstances produce modelling, particularly on human features, which is more sympathetic than that due to overhead lighting alone.

This is not an exhaustive list of the functions of a window, but we may summarise these four important aspects as view, release, variety and modelling. There is some value too in identifying how our visual expectation during the hours of daylight differs from that after dark. Briefly, we expect more light, of cooler colour quality, with more uniformity, and relatively high brightnesses on vertical surfaces (once again, four main points). Most of our experience of daytime visual conditions within buildings is in rooms with side windows, and this influences our response generally. We still broadly associate daylight hours with higher luminances—an overcast sky is likely to be five or ten times as bright as a luminous ceiling—and, traditionally, artificial light is warmer than daylight. In temperate latitudes, in areas not too far distant from an ocean, a cloudy sky is the commonest natural source; its very diffuse effect tends to reduce the variety of luminance—in contrast, despite the fluorescent lamp, our mental picture of spaces inside buildings at night is of a more punctuated effect associated with individual sources. Finally, the near-horizontal flow from side windows falls almost squarely on the opposing walls, and even with repeated interreflection the gross luminance pattern in typical daylit conditions differs characteristically from the afterdark situation, when ceiling and floor are relatively brighter.

There is, of course, no obligation to design electric schemes for daylight hours to produce related effects; but, whether we acknowledge them or not, these expectations constitute an important part of the human context of any scheme that we may develop.

Chapter 15

Form

The major significance of the development of the fluorescent lamp is that the cost of working illumination is now of the same order whether we provide it electrically or by daylighting. This can represent a revolutionary liberation from constraints on room geometry and hence on the total form of the building.

The effects can be illustrated, with some simplification, in terms of building forms supposedly typical of the past three decades.

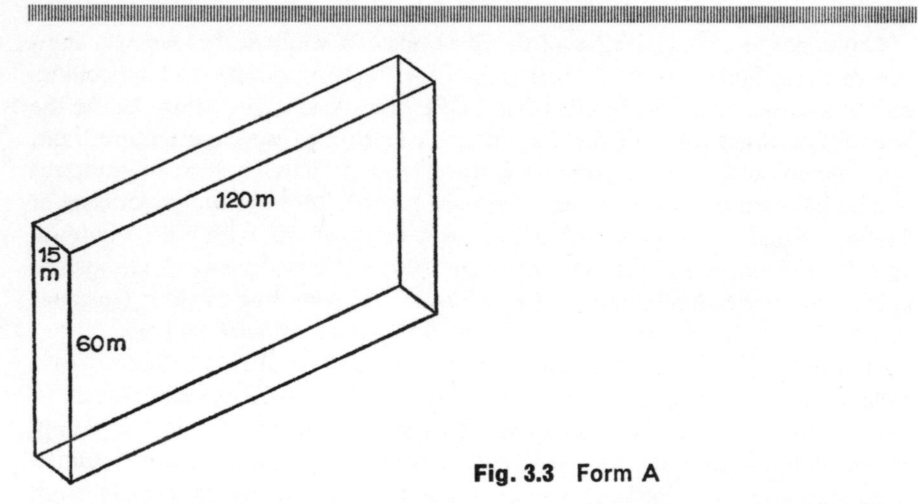

Fig. 3.3 Form A

A speculative office block built during the 1950s is likely to be a thin high slab. A traditional room depth of 6 or 7m on each side of a central corridor might give a thickness of 15m. Suppose the required accommodation is achieved in a building 60m high and 120m long, i.e. in a volume of 108000m^3 (form A, Fig. 3.3).

With the increased room depth associated with the p.s.a.l.i. approach, or some modification of it, the 1960s frequently saw a double corridor, or 'race-track', layout (Fig. 3.4).

Reading across the building from one face to the opposite one, the dimensions might be a 9 m room depth, a central zone of two corridors with no-daylight space between them, perhaps of 7 m, and a further 9 m room depth, to give a

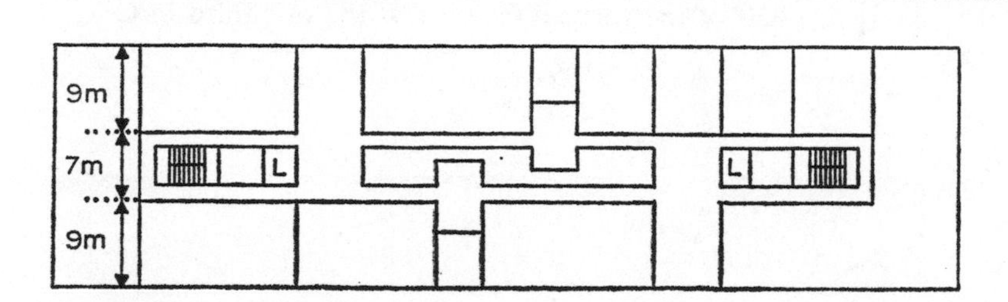

Fig. 3.4 Racetrack layout

total thickness of 25 m. Reducing both length and height, but achieving the same total volume, could mean a building 96 m long and 45 m high (form B, Fig. 3.5).

Once p.s.a.l.i. becomes p.a.l., it is possible to plan for a building with large open floor areas in a relatively low but deep form (form C, Fig. 3.6).

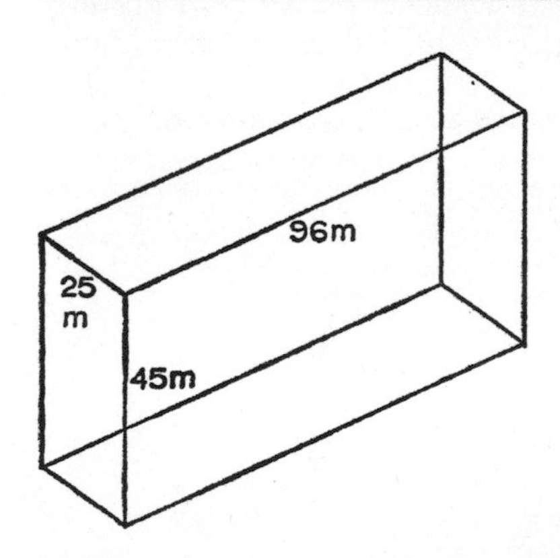

Fig. 3.5 Form B

This 1970s building has the same volume, 108 000 m^3, within a square-plan form and with a height of only half the horizontal dimension.

If the diagrams for forms A and C are superimposed, the contrast between them emerges strongly (Fig. 3.7).

In simple terms, C is half as long and half as high as A, but four times as thick. The surface areas of the three buildings (neglecting the undersides) are

A: 18 000 m^2
B: 13 290 m^2
C: 10 800 m^2

An even more marked contrast appears if we calculate the areas of glazing, assuming it to cover one half of the vertical faces in A and one-third in C.

A: 8100 m^2
C: 2400 m^2

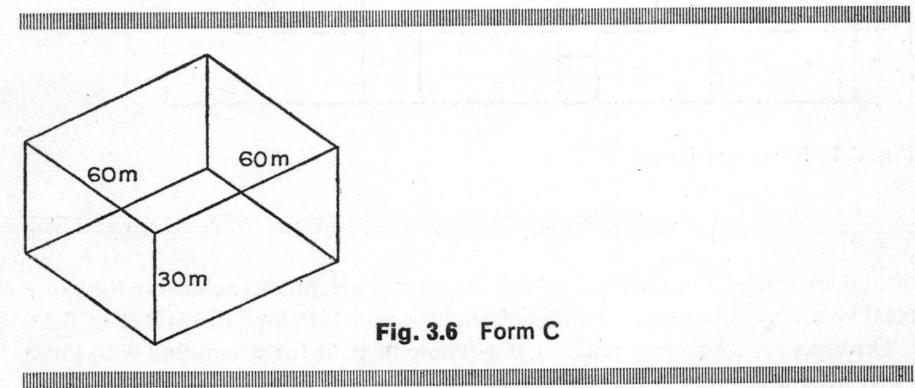

Fig. 3.6 Form C

Building A has essentially a low technology content; it utilises daylight and can have natural ventilation (though there are problems associated with opening windows in a building 60 m high). On the other hand, building C depends on

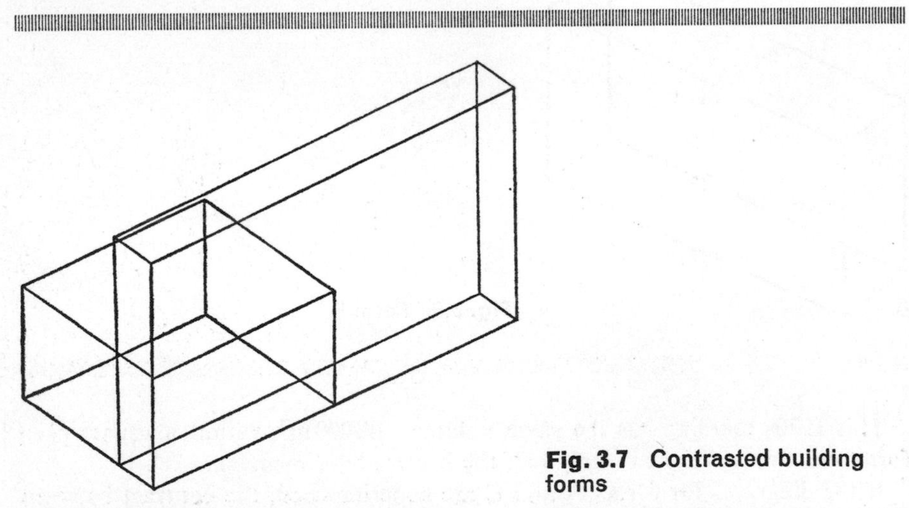

Fig. 3.7 Contrasted building forms

33

34

Plates 33 & 34 Cylindrical louvers in banking hall in Dusseldorf. Installation is unusual in that both diffusion and extraction of air occur above suspended ceiling

Plate 35 Integrated airconditioning and lighting in South-Western Electricity Board's Avonbank offices at Bristol

Plate 36 Airhandling fittings in a computer suite

37

Plates 37 & 38 Baffles combining visual screening and acoustic absorption

38

Plate 39 Fittings simply related to slatted ceiling bays in a school dining room

Plate 40 Hexagonal form on which Sunderland Civic Centre is based is expressed in lighting fittings

sophisticated services. Airconditioning is probably essential, and a sealed envelope likely. Electric lighting might represent 3% of the total initial cost of A, but lighting and airconditioning could easily account for 30% of the budget for C. This comparison is, of course, meaningless in itself, the only significant one being that of the total cost of each building with its value. Other elements in the cost can be estimated; C is obviously more economical than A structurally and in terms of cladding. Some aspects of value are quantifiable, such as the higher proportion of usable space in C; but there are others to which it is difficult to give any sort of weighting. For instance, in A, no working position is farther than 5 or 6m from a window wall; in C, some desks could be 20m or more from the perimeter. With the different proportions of glass in the façades, the less privileged worker in A sees much more of the outside world than his similarily placed colleague in C. This is a difference in value, but one difficult to express in currency. Again, an extended perimeter building can often be lighter and more sympathetic in impression than the squat box that economy suggests. It may be valuable to a commercial organisation to have its headquarters in the tallest building in town, a conspicuous landmark, but who is to say how much this is worth?

I have outlined just one example of a relationship between basic policy decisions on lighting and building form—and perhaps a rather naïve one at that. But the irresistible conclusion is that some decision on lighting strategy is among the first to be taken in any building. In many cases, the unspoken assumption may be that a traditional approach is being adopted; but the point may be made that speaking that assumption—i.e. formulating it as a conscious decision—at least demonstrates that the options have been considered.

Chapter 16

Fabric

The creative tension between the demands of structural stability and daylighting resulted in much that we enjoy in the buildings of previous centuries. The flying buttress is, classically, an engineering solution to the problem of larger apertures at the perimeter. The need for thickness in load-bearing walls led to the subtleties of splayed reveals and the gradation of brightness around mouldings and the like. But with steel, concrete, and curtain walling, the guidance that limitations represent was lost, and a whole new theory of window design has had to be developed from scratch. The irony of the situation is that windows were released from the obligation of providing illumination almost at the same moment, historically, that developments in structure allowed them at last to be any size or shape. The relationship between structure and daylighting is thus one of choice; if it exists, it is because we have chosen a form of construction which implies constraints.

We are concerned here, however, mainly with the relationship between electric lighting and structure, or building fabric generally, and we shall find that this appears in many ways.

The primary function of a structure is obviously to support the building; but the most important of its secondary functions is probably that of housing the services efficiently, economically, and—one hopes—elegantly (the three qualities go together in Utopia, and sometimes even in reality). We have, of course, seen over the past few years the imbalance that can follow the 'services sculpture' approach (an exquisitely managed run of conduit, for instance, terminating in a glaring batten fitting with a bare Warm White tube). Nevertheless, a thoughtful relationship between structural and services design makes one optimistic about the planning generally.

The physical integration of lighting equipment and building fabric can be pursued on a number of levels. It is most readily achieved in a building that hides its basic shell beneath shopfitting or something equivalent. Most recessed lighting fittings are designed on the assumption that they will be mounted in suspended ceilings. Preparing the fabric for the lighting hardware is thus a matter of cutting a hole in a tile rather than casting one in a slab. The tolerances in fittings manufacture are relatively fine—those of the factory rather than the building site.

With increasing standardisation, components may no longer need cutting and fitting. Recessed fluorescent fittings are often sized in relation to the 300 mm and 600 mm modules adopted by ceilings manufacturers: two, three or four 40 W tubes in a rectangle 1200 mm × 600 mm, and so on. Unfortunately, the internationally accepted range of tube lengths is not completely ideal in this sense (Reference 2.36). In most cases, the overall length of a fitting (above the ceiling plane) is greater than the nearest modular length. The 40 W tube, for instance, needs a fitting about 1240 mm long, which obviously makes continuous mounting difficult. It is possible to produce a fitting for 85 W lamps of nominally 1800 mm length without exceeding 1800 mm overall, but this is a tight fit. It may be necessary to make the fitting the next larger modular size: the 65 W tube (nominally 1500 mm) is often housed in fittings that replace three 600 mm square ceiling tiles.

Another approach to the physical integration of recessed fluorescent lighting is to form some sort of trough in builder's work and to mount tubes within it. Overlapping them can achieve two ends simultaneously: the avoidance of dark patches where tubes meet, and reconciling tube dimensions with the length to be covered.

Fig. 3.8 Arrangement of tubes within trough

The trough may also be assembled from manufactured components. Some systems of this sort offer a choice of light-control media and facilities for acoustic insulation where a trough passes over a partition. Decisions on switching necessarily tend to limit flexibility.

There exist very extensive ranges of fittings for incandescent lamps intended primarily for recessing in suspended ceilings. Their feasibility depends on the recess depth available, but the space the fitting needs varies considerably from type to type, so the necessary dimensions should be established at once. A 'black hole' downlight for a vertical reflector lamp may need 300 mm depth or more, but less than half this suffices a fitting for a horizontally mounted g.l.s. lamp. Fluorescent fittings can sometimes be accommodated in 100 mm (Fig. 3.11).

These are largely mechanical matters, but reconciling lighting hardware with the fabric of a building is more than a question of clearances.

Lamps that are symmetrical about a vertical axis tend to generate fittings that are circular in plan. In contrast, linear sources such as fluorescent tubes produce rectangles. Depending on the circumstances, either form may represent a visual discord. The tradition of circular fittings is now a long-standing one, and surface or pendant fittings of this shape rarely present problems. But recessing a circular downlight into a boarded ceiling may prove clumsy, and needs perhaps a special square mounting with a side equal to a whole number of board widths. An alternative is to design the fitting as a square housing for the lamp, and a number of manufacturers have produced ranges of 'square cylinders' over the past few years. The disadvantages include the use of more material, production problems in extrusion of the larger sizes, and the fact that the cutoff along a diagonal of the square is less than that along a line parallel to a side.

Recessed fittings for incandescent lamps are produced for use with 'wet' or 'dry' ceilings. Those for 100 W or 150 W ratings usually show no dimensional relationship to preferred modules, being made as small as is consistent with their easy handling and with thermal needs. But recessed fittings for higher-wattage incandescent lamps or for discharge lamps are now appearing in the 600 mm-square size (recent improvements in colour-rendering resulting from new phosphors in fluorescent-bulb mercury types are leading to their being advocated for commercial interiors where hitherto they would have been thought unsuitable). A square fitting of this type happily takes as its light-control medium a panel of small-cell square louver, a square stepped lens, or a panel of non-directional prismatic plastics—or a plain opal panel if that is appropriate, though the source brightness suggests that more sophisticated glare control is desirable.

The rectangular fluorescent fitting that is happily at home in a room of conventional shape may be difficult to place convincingly in a space with walls that are curved or not perpendicular to each other. Circular tubes were developed to meet the demand for a round fitting in the early days of fluorescent lighting, but they have a limited usefulness, particularly if a generous illumination is required (relatively high tube-replacement costs also restrict their application).

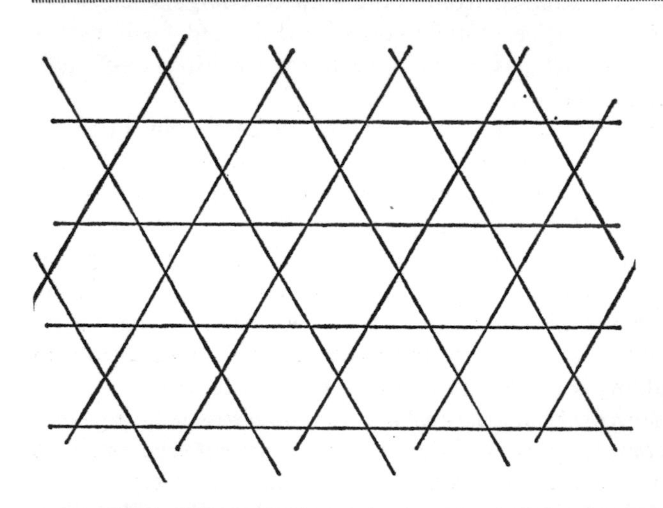

Fig. 3.9 Shapes produced by lines parallel to three axes at 120° to each other

Tubes with other special shapes—'U' or 'W'—may have advantages, but they tend to produce rectangular fittings.

Most buildings derive from two axes at right angles. In some, the one or the other is the major axis, or may arbitrarily be declared so. Running all linear fittings in the building parallel to this will provide a discipline that can read convincingly through all the spaces, however irregular they may be. Some recent structures are based on three axes at 120° to each other, but, even here, making one of these directions dominant may be effective. Three axes of this type produce equilateral triangles and regular hexagons (Fig. 3.9).

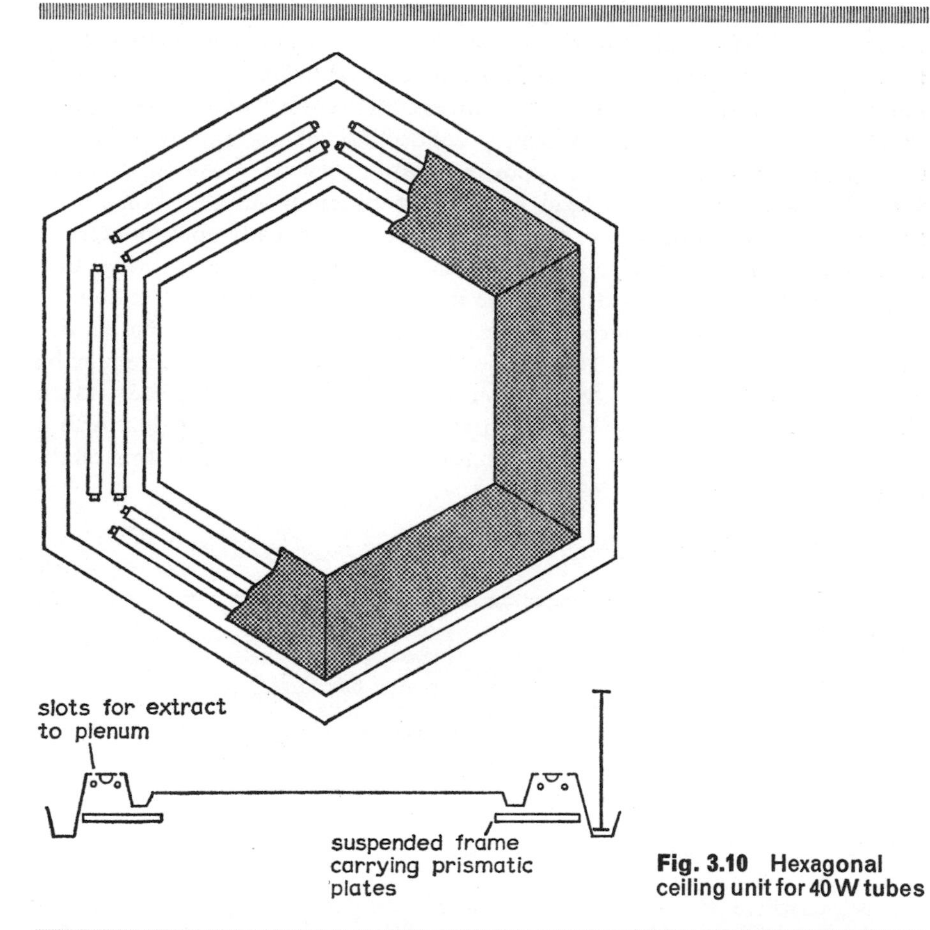

Fig. 3.10 Hexagonal ceiling unit for 40 W tubes

A hexagonal plan shape is quite suitable for lighting fittings, though few catalogued examples currently exist. The scale might vary from a 'hexagonal cylinder', housing a reflector lamp, to a hexagonal unit in a large ceiling with straight tubes along each side (Fig. 3.10).

A circular tube (or a pair, 40W and 32W) may readily be fitted into a hexagonal housing, and hexagonal-cell louvers are available as standard.

All these are devices that may or may not prove relevant to a particular case; but, when satisfactory physical integration with a building fabric is not possible using existing lighting hardware, some special design work is called for—and, because it is special, generalisation about it is difficult.

One basic principle, however, is to associate lighting equipment with the flesh of the building rather than its bones. When the bones are completely hidden, we have the situation as it appears, say, in many department stores; both incandescent and flourescent equipment is, as we have seen, happily accepted by the shopfitting. Imagine, however, a different situation: a room with walls in fairfaced blockwork but a suspended ceiling of acoustic tiles. Then these tiles are the right part of the fabric to associate with the lighting equipment, because their status is similar—both are 'fitting-out' items applied within the basic shell. That this is not simply connecting lighting and the ceiling is confirmed by a reference to the opposite situation: part of a church, perhaps, with stonework tracery overhead and timber panelling on the walls. Then sources concealed at the top of this panelling seem essentially appropriate.

Suspended ceilings provide the opportunity of housing recessed fittings that we often need, although by itself this need is scarcely sufficient justification for such ceilings. Their main purpose is often to conceal airducts and other service runs, and tolerant liaison is called for so that all the demands for space may be met in the minimum depth. Separation in plan is frequently the ideal to be aimed at, though there may be enough vertical space for a minor duct with a shallow lighting fitting beneath it within the depth of the major duct that establishes the plenum size (Fig. 3.11).

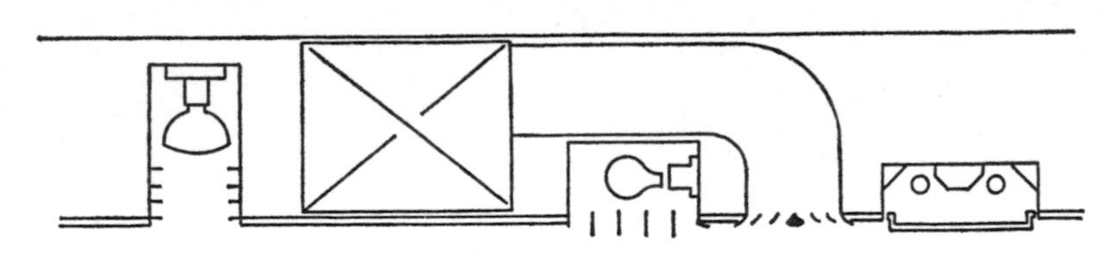

Fig. 3.11 Utilisation of plenum depth

The gap where basic structure and applied fabric meet can provide mounting positions for lamps. Probably the most common example is the suspended ceiling that stops short of the wall or a beam (Fig. 3.12).

There are many possible variations of this detail. It may be applied even with a curved perimeter, provided that the radius is not too small—the appropriate length of tube will naturally vary with the curvature and the extent of overlapping (Fig. 3.13).

Built-in lighting of this kind implies the alternatives of complete batten fittings or tubes in spring clips with remote gear. The former are quicker and easier to install, the latter more flexible. Control gear is best attached to firm basic structure, to avoid acoustic amplification of any hum, so the loading on the ceiling is reduced.

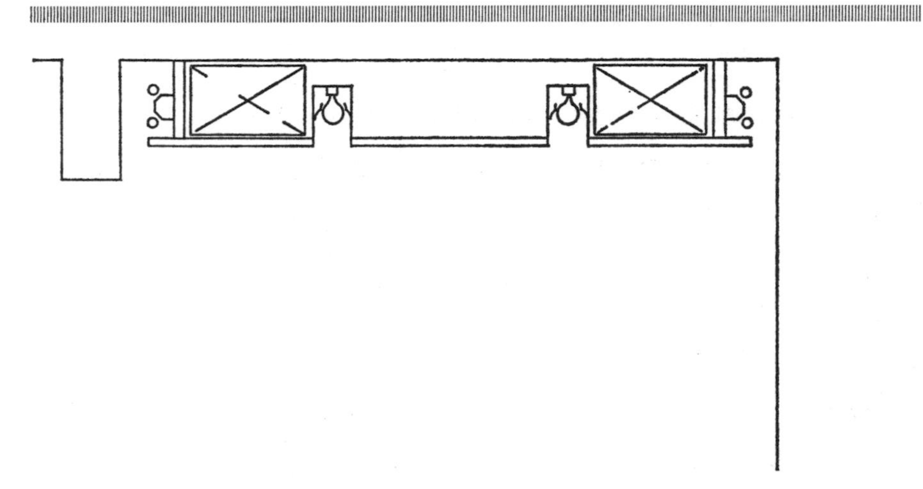

Fig. 3.12 Integration of lighting and fabric

Another advantage of the scheme that uses tungsten fittings and fluorescent tubes concealed in a feature of the room fabric is that the latter may be switched off without creating the impression of lamp failure—valuable where one wants to use both components by day and the incandescent lamps alone after dark.

The association of lighting equipment with shopfitting and suspended

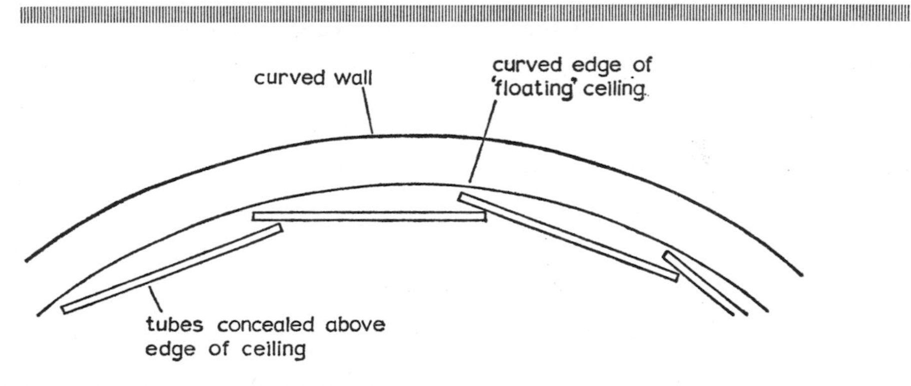

Fig. 3.13 Straight tubes following curve

ceilings throws into relief the problems in buildings with little or no applied fabric. Where the basic structure is not merely revealed but celebrated, lighting fittings are in danger of seeming incompatible with it. There are certainly inconsistencies in trying to cast holes in the slab, say, to take an array of recessed fittings. On one level, this is to condemn the building for its life to the lighting technology available now. On another, it is to neglect the fact that the thermal design of the fitting almost certainly supposes some free air around the lamp housing.

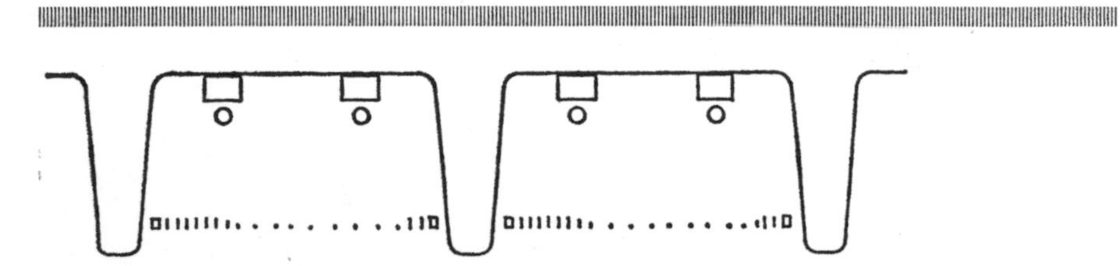

Fig. 3.14 Panels of louver reading as secondary ceiling

One solution is a frank acceptance that the lighting equipment is applied to the surface. In the brutal extreme, conduit or trunking may run on the surface too. The danger is that, even if the layout is planned meticulously in advance, the result may look like an afterthought, as though we were back in the bad old days when the building was effectively finished before lighting was considered. Another possibility is to conceal light sources in troughs or other spaces that the structure spontaneously provides, or can be manipulated to offer without loss of apparent spontaneity. The multiple column, the double beam and the coffered slab are all valuable here. Sometimes, the lighting equipment resembles applied fabric, as with a suspended 'floating' panel of louver within a coffer (Fig. 3.14).

Depressions cast in the underside of a slab can take surface fittings to produce an intended impression without unduly limiting future modification (Fig. 3.15).

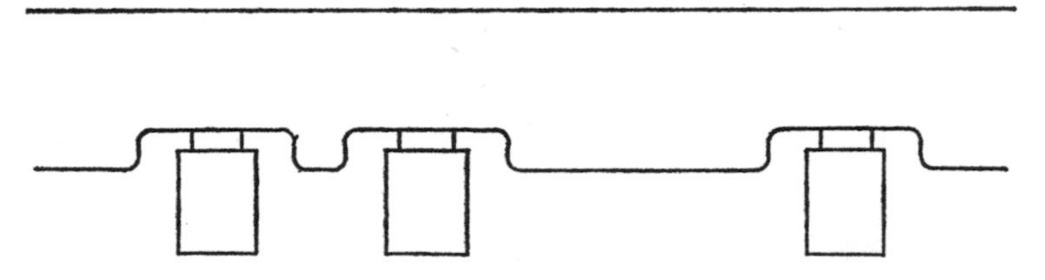

Fig. 3.15 Depressions in structure to take lighting equipment

They must be oversize in plan, but may be quite shallow while still making their point.

These various methods of reconciling lighting equipment with revealed basic structure often fall into place if fluorescent tubes and tungsten lamps are treated in different ways, the former positioned in the crevices provided by the structural engineer (possibly with some prompting) and the latter explicitly applied to the surface, either individually or on runs of track. Such an approach can form the basis of a coherent visual style.

Although it can hardly be regarded as building fabric, furniture provides opportunities for mounting lamps, and it is legitimate to depend on using it either for local direct lighting or for upward light. The latter technique is sometimes useful in relighting an existing interior—particularly a high one, where it may avoid problems of wiring and maintenance. In a new structure, it may be more appropriate to provide the general building lighting from sources apparently in with the bricks whereas specific local working needs are met from the furnishing. This automatically takes care of change-of-use problems; so we may see it as consistent with a 'loose fit' philosophy. Any lighting in furniture naturally depends on an electricity supply being conveniently and safely available (the increasing use of carpet tiles helps here). Greater flexibility in the management of lighting implies additional responsibilities for the user, which he may welcome, but which will also mean extending the instruction and maintenance manual that comes with the building.

The discussion in this chapter has been largely concerned with the physical integration of lighting hardware and building fabric. The hardware is, however, only a means to an end, namely the lighting effect. It is how the building is lighted, rather than how the fittings are mounted, that is of the essence in lighting design.

The structure of a building, of course, has major acoustic and thermal implications. How these are related to lighting should emerge from what follows.

Chapter 17

Sound

Incandescent lamps are usually completely silent in operation. Occasionally, a vibration of audible frequency may be set up in a filament shortly before it fails, and the lamp—like a swan—will sing before it dies.

Although it may be unnoticeable, there is some noise from inductive control gear for any discharge lighting, including fluorescent tubes. We have all encountered situations where it appears as an unpleasant hum, or worse; so it is desirable to be aware of the main factors that influence it.

A ballast 40mm × 30mm in cross-sectional area is not necessarily noisier than one measuring 60mm × 40mm, but the probability is that it will be. There is thus more likely to be a noise problem with slim-sectioned economy batten fittings than with a better-quality range, from the same manufacturer, with a more substantial shape.

Whether noise from lighting equipment is noticeable depends to a major extent on the ambient noise level. A fitting that was perfectly satisfactory in a city office could prove not to be so in a country hospital.

Difficulties can arise because a slight hum from gear is amplified mechanically. An area of suspended ceiling or a shopfitting panel may act as a sounding-board, and a fitting or some separate gear that was quite acceptable on the bench could result in irritation or distraction. The answer is to attach the gear to some firm part of the fabric, and this may influence the choice between a complete batten fitting and tubes mounted in spring clips. Where quiet conditions are sought, all the control gear may be moved to another space in acoustic isolation.

A mechanical strain in the metalwork of a fitting can cause the vibrations transmitted from the choke to produce excessive noise. This may be the case when a 2·4m batten is screwed firmly to a ceiling that is not quite flat, in relighting an older building perhaps. One-half of a turn of a screw may make all the difference, but searching for this sort of correction is tedious and frustrating; the need for it might be avoided by thought at the design stage.

Noise from lighting equipment cannot be called a major problem, but there are dangers in the comforting belief that nobody really bothers much about it, and these are somehow accentuated by the differences between laboratory conditions, under which any testing is carried out, and the facts in the field. Experience seems to confirm that 1·2m tubes should be considered when quiet operation is specially desirable.

A high-frequency supply for fluorescent lighting would not be advocated solely on acoustic grounds, but, if the frequency were high enough, it could move the 'noise' out of the audible range (the other advantages include freedom from flicker, and high-efficiency, lightweight, capacitive control gear).

The control of room acoustics depends largely on the provision of surface areas with reflective or absorptive characteristics—indeed, in many cases, the recipe is the simplest: as much absorption as possible. Baffles and the like that increase the area available for the application of absorptive finishes may also act as screening for light sources. Some European examples of extensive landscaped offices have a visual ceiling made up of what is in effect a large-scale egg-box louver with acoustic treatment; above this simple batten fittings are mounted on the structural ceiling. There are proprietary or packaged versions of this idea. Most lighting equipment itself has hard sound-reflecting surfaces; indeed, the proportion of the ceiling taken up by a layout of recessed fluorescent fittings may worry the acoustics adviser. Many metal louvers transmit sound freely and permit the application of acoustic treatment to the slab above them. At least one system of diffusing panels for a luminous ceiling claims absorption characteristics comparable with conventional ceiling tiles.

In many modern buildings, a major sound problem appears in the lack of privacy between offices on opposite sides of a partition that finishes at suspended ceiling level. This acoustic leak can be made worse by lighting fittings that puncture even the modest barrier that the ceiling represents. Sometimes when a continuous trough runs through a large area, and partitioning is subsequently brought up to its underside, this equipment provides an immediate and direct path for sound. There is no magic solution to these problems, but recognising them must influence planning generally. In large open office spaces, the problem of acoustic privacy is at least immediately apparent. One solution suggested is 'acoustic perfume', the provision or acceptance of a moderate level of unobjectionable background noise, which implies a maximum distance over which face-to-face or telephone conversations may be heard. A level of 50–55dB has been recommended (W. Zeller, quoted in Reference 2.37). Such background noise may come from the airconditioning system, and it might seem strange in these circumstances to worry about a minor hum from lighting control gear. Anything intermittent is to be avoided—in fact, we come back to the difficult but basic question of what noise annoys.

These then are some instances of direct relationships between lighting and acoustics that may influence building design; but the more important interactions are indirect, and at the strategic level. An example or two may be given briefly.

The viability of office buildings in city centres has been questioned because of the problems produced by traffic noise (Reference 2.38). One possible solution is to make the core into a shell—in other words, to produce an acoustic buffer zone between outside and inside and to use this for circulation, services and so on. This would mean working areas remote from the exterior, and made possible by electric lighting, but with their human acceptability still open to question. If the idea is pursued, long views and some reflection of natural change in the visual conditions seem essential.

The mass of the structure has major implications in terms of acoustic insulation and also a profound effect on the building as an energy system. Since the major energy input may well be represented by the lighting, this constitutes a fundamental, if complex, link between decisions on sound and on light.

Acoustics and lighting are both affected by most of the strategic decisions in building design: the technological emphasis; the sophistication of the services; the size of typical spaces; whether the windows are large or small, sealed or openable. These are the things that make the building what it is.

Chapter 18

Heat

Suppose a 100W lamp hangs from a flex in the middle of a room without windows. When it is switched on, an electricity supply is made available, and electrical energy is consumed at the rate of 100W (i.e. 100 joules of energy per second). What happens to it? The passage of the current through the filament turns it into heat, but, since the filament rapidly settles down to its steady operating temperature, it must be losing energy at the same rate as it is being supplied with electrically. Part of this loss is by conduction and convection (through the filament supports and the lamp gas filling) and part by radiation. Of the radiation, some is infrared—invisible heat radiation—and some is within the visible spectrum, i.e. light (there is also a small quantity of invisible ultraviolet radiation outside the other end of the spectrum). The major part of the energy supplied is thus directly converted into heat, either radiant or convected and conducted. The remainder is light, but the amount of light in the space does not increase, although the lamp stays switched on, so it must be absorbed in the room surfaces at the rate it is being produced. In being so absorbed, it raises the temperature of the surfaces; so it too is converted into heat.

The 100W rating of the lamp thus expresses both the rate of supply of electrical energy and the rate of production of heat (in SI units, both are in 'watts'). The room is made warmer than its surroundings, and so begins to lose heat to them. Its temperature rises, and the heat flow through the boundaries of the enclosure increases in proportion to the temperature difference between inside and outside. When this heat loss rate rises to 100W, the point of balance is reached. The steady temperature that results depends, of course, on the size of the room and its thermal characteristics, or more specifically the areas of the surfaces and their conductance. In a small, well insulated room, even a 100W supply of heat—the lamp—can produce a significant temperature rise.

There is nothing mysterious about heat from light. Indeed there is no artificial source of light that does not produce heat, and any electric lamp consuming x watts represents a heat source of x watts. It is true that, in a room with a window or an open door, the light and radiant heat that escape through these openings are absorbed in exterior surfaces and so do not contribute directly to the temperature rise within the enclosure. But this is a marginal effect, and can be ignored for most calculation purposes.

Heat from lamps has sometimes been regarded merely as an embarrassment, but a more positive approach is to think of it as potentially useful energy; useful, that is, if we can find how to manage it.

Let us first recall the way heat is dissipated by different types of lamp (Chapter 11). More than half of the energy supplied to a normal incandescent lamp appears as radiant heat. Typical figures are

radiation	60%
convection and conduction	30%
light	10%

There will be minor variations with the size and rating of the lamp, as there will be with fluorescent tubes, where the additional factor of heat dissipation from the control gear should be taken into account. Typical figures here are

Heat from control gear:	
mainly conduction and convection	20%
Heat from lamp:	
conduction and convection	40%
radiation	25%
Light	15%

Of 100 W supplied to an incandescent lamp, then, about 10 W is converted to light. Much of this is at the red end of the spectrum, where the eye is relatively insensitive. The 15 W of power converted to light from 100 W supplied to a typical tube represents about three times as much light, since a large proportion of it appears in the yellow and green parts of the spectrum, where sensitivity is much higher. The precise relationship depends on the tube colour, but, for equal quantities of light, we may make the following comparison in representative figures:

	Tungsten	Fluorescent
Input	300 W	100 W
Convected and conducted heat	90 W	60 W
Radiant heat	180 W	25 W

These data refer to energy dissipation from the bare lamp. When the lamp is housed, as it normally is, within a fitting, some of the radiation warms the reflector, the louver, or other components, and the pattern of heat loss from the complete system of lamp and fitting will be different—probably less radiation and more loss by convection and conduction. This is an important distinction, because the radiant heat goes wherever the light goes, and the heat that might be removed by, say, a flow of air is limited to the convected and conducted part. For a fluorescent fitting recessed into a suspended ceiling, however, this represents the major part of the energy: the proportion of heat that can be extracted by a flow of air through the fitting into the space above the ceiling can be as high as three-quarters (Fig. 3.16).

Airhandling fittings for incandescent lamps are uncommon, largely because the proportion of the energy that can be carried away is so much smaller. They may, however, be used for other reasons, such as the wish to limit the

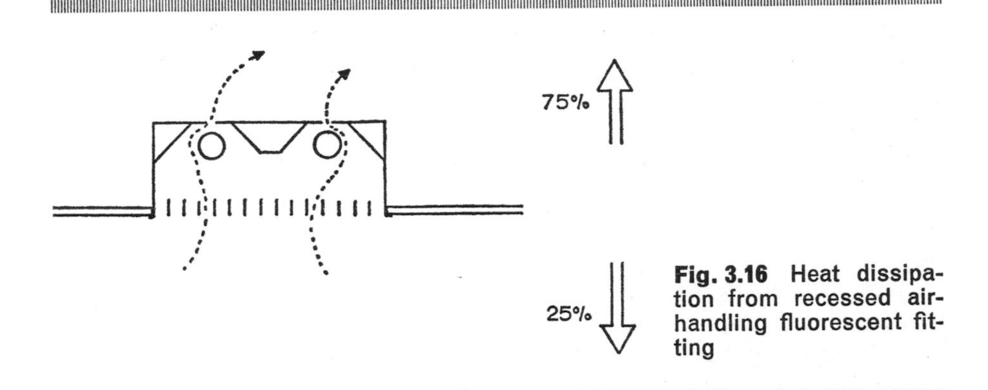

Fig. 3.16 Heat dissipation from recessed air-handling fluorescent fitting

number of separate pieces of hardware in a ceiling. Some industrial fittings for high-wattage tungsten or discharge lamps utilise the convection airflow for its so-called 'scavenging' effect in reducing the deposit of dust on the upper surfaces of the lamp.

The exception to the principle mentioned above—that infrared-heat radiation must accompany the light from a lamp—occurs in types with a special 'dichroic' reflector (Fig. 3.17). This acts as a filter, in the sense that it reflects most of the visible light and transmits some of the radiant heat.

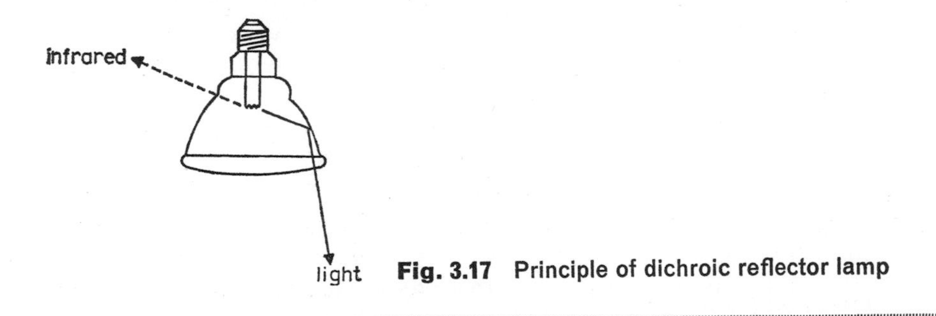

Fig. 3.17 Principle of dichroic reflector lamp

The separation is not complete: some light is lost through the reflector, and radiant heat is still noticeable in the forward direction; but there is certainly less sensible heat at a given distance from a dichroic spot of 150 W than from a corresponding 100 W lamp with a conventional metallised reflector. The heat that is radiated upwards from a lamp in a suspended ceiling tends to warm the structural slab above, and most of this form of heat can be removed by a flow of air through the plenum.

If we turn now to consider what proportion of the energy needs of a building may be represented by the lighting, we realise this is like asking the length of a piece of string. The heat from the lamps may be an almost negligible part of the total requirement, or it may be the major part. In an exposed detached house with large windows and with walls and roof of traditional construction, turning on the lights in January will make little difference to the load on the boiler. But a deep-plan office building with restricted areas of double glazing and opaque cladding of low thermal transmittance is a different situation. With the incidental gains from people and machines, the lighting load needed for illumination levels in the 1000 lux region is sufficient to maintain comfortable temperatures inside with no additional source of heat.

When the situation is discussed on this rather naïve level, it seems at first that any energy supply can be sufficient provided that the insulation is thick enough. This might be true if there were no need to ventilate the building, but the necessary flow of air through occupied spaces results in 'ventilation loss' to be added to 'fabric loss'. In a typical domestic situation, air changes according to the book might account for 20–30 % of the total heat loss, but the more restricted ventilation of practice often means a lower figure. It is clear, however, that, as insulation improves, so ventilation loss comes to represent a larger proportion of the total heat loss—in a modern office block, it could be over 50 %. In that case, improving the insulation further shows diminishing returns, and attempting to make use of lighting heat seems difficult, as it is being removed by air changes more rapidly than it escapes through the fabric.

The answer is to recirculate most of the air, treating it along the way and introducing the smallest acceptable proportion of fresh air from outside. But airconditioning has a much greater relevance to the use of lighting heat than simply that of reconciling total heat loss with total energy gain: it provides, in addition, a medium for the management of the heat, for its redistribution from areas of excess to those of shortage (such as the inner zone and the perimeter, respectively, in a deep building).

The integration of lighting and airconditioning may bring together a number of largely independent advantages on different levels of sophistication.

Visual coordination is the first. Simply lining up airhandling apertures and lighting equipment improves the appearance of a ceiling, particularly if the hardware is dimensionally related. This is taken one stage further in the airhandling lighting fitting, whatever its other advantages, and installation work is simplified.

One of the most important, if apparently incidental, advantages of extracting room air past the tubes in a dual-purpose fitting is that it provides a way of keeping the temperature of the lamps close to the optimum (Fig. 3.18).

This is valuable, because the fluorescent lamp is temperature-sensitive. Most

Plate 41 In a situation such as this, attempting to recess fittings into primary structure would lead to many difficulties, and suspended fittings would be entirely out of keeping. Luminous objects at floor level bring welcome vitality to scene

Plate 42 Reflection in specular finish to exposed structure adds visual excitement to approach to auditorium at Paramount cinema, London, and avoids dullness often associated with predominantly indirect lighting

Plate 43 Radio and t.v. department, Globus, Zurich. Cubic ceiling hangs from power-supply track, with each lighted cube housing four 20 W tubes above louvers

Plate 44 Café in C & A store, Amsterdam. Recessed fittings over bar area, but pendants and wall brackets elsewhere furnish space and establish its character

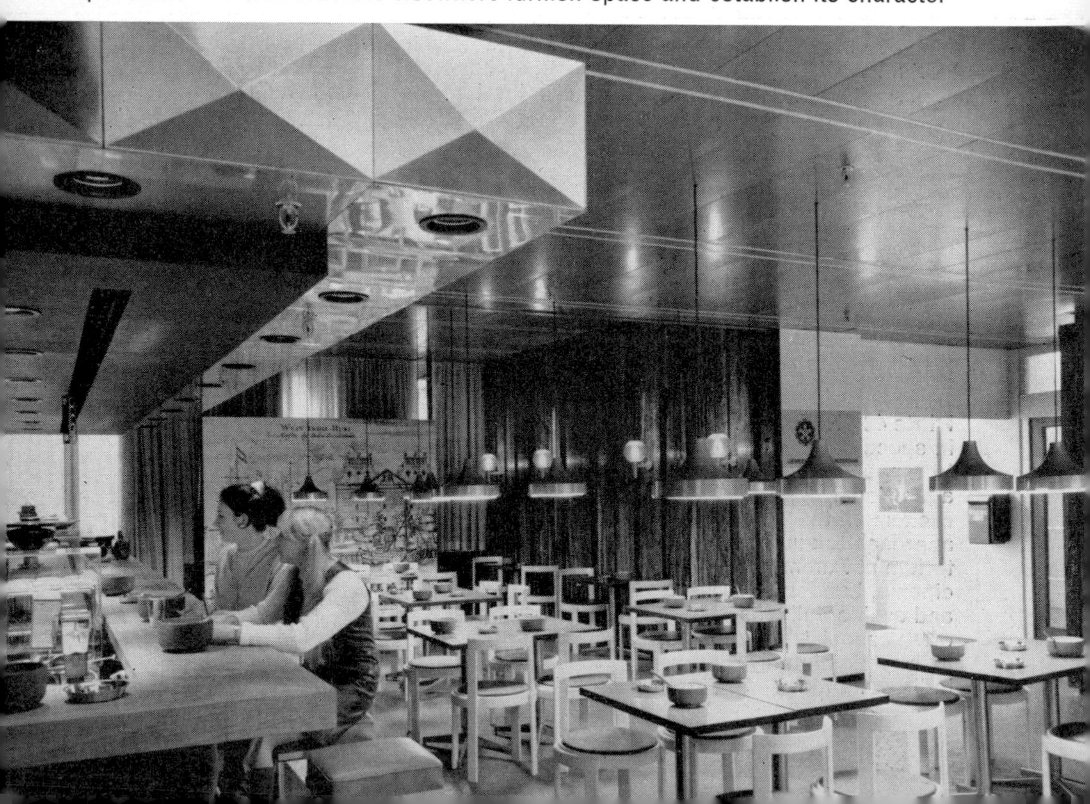

Plates 45 & 46
Cylinders as theme in shopfitting design: lighting fitting develops into display prop

45

46

47

Plates 47 & 48 Two sections of main arcade through Woluwé Shopping Centre, Brussels

48

types are designed for operation in free air at 25°C. In an enclosed fitting, the higher ambient temperature can mean a loss of 10–15% in output. In a large installation, nine airhandling fittings can give more light than ten unventilated types, with economies both initially and in running costs; on an immediate cost level, this is perhaps the most telling argument for an integrated system.

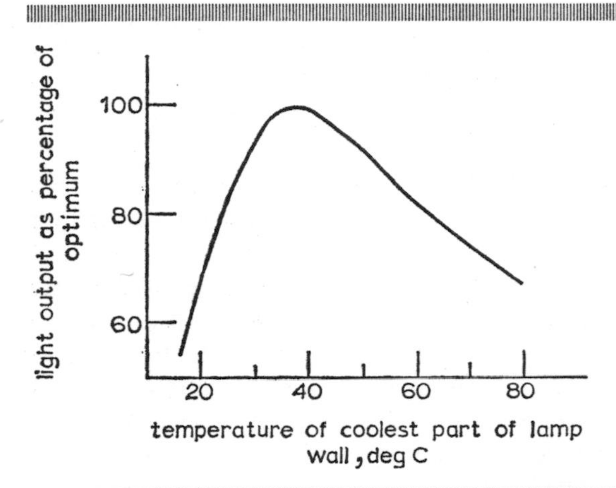

Fig. 3.18 Relationship between light output and tube-wall temperature

It used to be thought that the flow of exhaust air past the tubes themselves would lead to soiling, but experience has shown that depreciation due to dirt is less than in corresponding traditional fittings. There are many alternative modes of operation, the fittings being used either for extraction or for introduction of air—or both—and ducts carrying exhaust and conditioned air, or the void above a false ceiling functioning as a sealed plenum for one of them. But the system now most widely used (Fig. 3.19) employs independent apertures for the air supply; they are fed by insulated ducts passing through the plenum, which is at a negative pressure to stimulate the flow of return air over the tubes and through slots in the reflector.

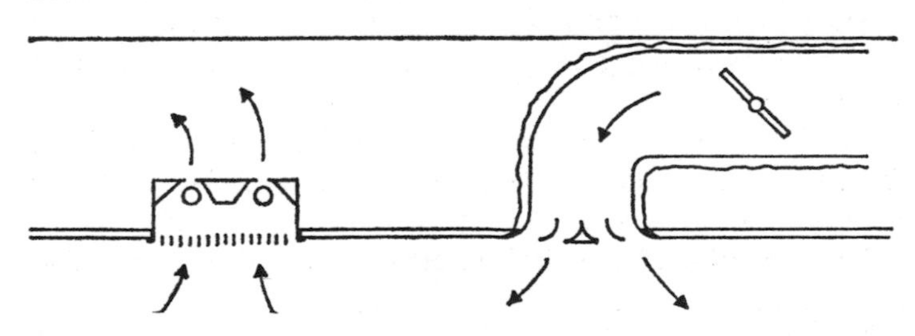

Fig. 3.19 Supply and extraction of air

This arrangement permits the inlet apertures to be designed for the best injection pattern of conditioned air into the space; the detail of openings for extraction is much less critical, and so easier to incorporate in the lighting fitting. If the light-control medium is some form of louver, the air can pass through with little obstruction; otherwise—if, say, a prismatic plate is used—slots along the fitting or across the ends are introduced.

Even with the details of airhandling at the ceiling sorted out, there remain many ways in which the airconditioning system can achieve its traditional objectives, and the relatively recent one of heat recovery—specialist advice is obviously called for. Our present purpose may be served by giving, purely as an example, a brief description of one system.

At its heart is the refrigeration machine based on a recurring cycles of changes in a medium such as ammonia. Energy is supplied via an electric motor that drives a compressor causing the medium to circulate through evaporator and condenser. In the former, heat is taken in, and, at the latter, given out. Water running through cooled pipes around the evaporator is thus chilled, and the heat developed at the compressor may also be taken by piped water to a cooling tower where a flow of atmospheric air carries it away. Air is introduced to each room through a number of induction units, at the perimeter beneath windows or spread over the ceiling. Its emergence from nozzles stimulates a flow of room air through the unit over the coiled pipes that are carrying the chilled water from the evaporator—if cooling is required. If heating is needed, this chilled water is turned off, and the air entering the room is heated before it reaches the induction unit. Any type of fuel may be used for this.

In temperate latitudes, it is often the case that cooling is needed in one part of a building and heating elsewhere. In a deep office block, for example, cooling will always be required near the core, and heating for much of the year at the perimeter. With conventional airconditioning, energy is being used to extract heat and take it to the cooling tower for dissipation, while, independently, more energy is needed to provide the heat for the air entering the spaces where warmth is required. This is essentially wasteful.

The heat-recovery adaptation of the system involves only minor changes, and can be achieved with standard equipment. The essential difference is that there is a twin condenser. Heat from one part takes the place of the independent source, and is applied to warm the returning air where necessary. The other part carries to the cooling tower only energy that is in excess of the total needs of the building at the time.

The extraction of air from the occupied spaces is through the lighting fittings, so that the heat from the lamps is largely removed at source, for recycling or dissipation as is appropriate.

The local control of the system—the choice between warmth or 'coolth'—depends on room thermostats. Systems of this sort have been used to maintain an almost steady 21°C internally while outside temperatures fall as low as —4°C. Such a balance usually depends on a lighting loading in the main working areas of about 40 W/m^2.

The relative size of the components in any expression of the energy balance in a building will depend on its site, orientation, form, glazing and so on; but rounded figures for a fairly conventional office building for 1000 occupants might read

	Gain kW	Loss kW
Fabric loss		300
Energy to heat fresh air coming in		400
Heat recovered from exhaust air	140	
Heat from lighting	350	
Heat from occupants	90	
Energy from compressor	170	
Energy from pumps and fans	50	
Balance to cooling tower		100
Total	800	800

The 'balance to cooling tower' represents a safety factor, or help during the 'topping up' period after the building has been empty. The extent to which the temperature drops overnight, or during a winter week-end, depends on the heat storage represented by the fabric of the building, i.e. on its thermal capacity, which is directly related to the weight of the structure (it is significant how often the question of structural mass has arisen in this discussion of aspects of building design). Additional thermal storage can be provided in the form of large, well insulated water tanks, which are charged by any spare heat before it is diverted to the cooling tower. Circulation of this warm water through what are normally the chilled water coils of the perimeter induction units makes them act as simple convectors for, say, one or two hours before work begins on a Monday morning.

The airconditioning system will naturally include other normal features such as humidification and filtration to remove odours.

A great deal has been published on heat recovery over the past few years. The UK Electricity Council and the area boards have produced booklets on a number of important particular buildings (References 2.39–2.41). A general survey, under the title 'Light and the total energy input to a building' and mildly academic in tone, appeared in *Light and Lighting* for September 1970, in the form of a translation of Prof. H. W. Bodmann's inaugural address at the Lighting Engineering Institute, University of Karlsruhe, W. Germany (Reference 2,42). Another survey of importance, with a full list of references, appears in Prof. J. K. Page's paper to the 1970 IES national lighting conference at York, England (Reference 2.43).

Chapter 19

Systematic lighting design

Apart from any question of practical convenience, there is an intellectual fascination about reducing things to a system. Computer programming and programmed learning have recently tended to emphasise the value of analysing a process into small sequential steps. This not only makes assimilation easier, but also increases our understanding. It is against this background that attempts at formulating systematic techniques for the design of interior lighting have been made.

It is clear at once that no simple linear program can be adequate. No series of prescribed steps in a rigid order can accommodate the complexity of the problem. Something more flexible, however, seems promising. Any feasible system must allow for interaction between requirements, for personal estimation of the weighting of criteria. It must provide loops and feedback and similar facilities beloved of the computerate.

An approach proposed by R. T. Dorsey, of the General Electric Co., Cleveland, Ohio, USA, seems to meet this need. It was formulated at the 1971 conference in Barcelona, Spain, of the Commission Internationale de l'Éclairage (CIE): 'A unified system for the esthetic and engineering approaches to lighting.' Its declared purpose was 'to outline a framework to bring the parameters of environmental design together in an organised flow from problem to solution... The starting-point is the situation for which the designer must provide a solution. It will be seen that there are three design paths—one related to the space, one to the psychology involved, and one to the functions.' This refers to a 'design procedure network', a diagram crystallising the method and made up of balloons labelled 'the situation', 'the psychology' and so on, with arrows leading from one to the next or indicating a 2-way exchange between them. The design balloons resulting from the three paths specified above lead to a daunting final balloon labelled 'relate, evaluate'. But it is surely in relating and evaluating that the real problems arise in lighting design, or many other kinds of design for that matter. The network is developed into many subsidiary diagrams. One titled 'Evaluation phase' reminds us that cost is related to wages, to 'behavioral benefits', to value added, to safety and so on; it also indicates that energy is a link in the integration with thermal and acoustic design.

Any attempt at summarising a closely argued paper must mean distortion, and my random comments here prove nothing. The interested reader will find the paper by Dorsey very effectively presented, with generous illustration, in a special issue of the *International Lighting Review* (Reference 2.44). I can only add that I find the paper extremely interesting, but ultimately unconvincing in its declared purpose.

A completely different approach is evident in Peter Jay's paper, 'Inter-relationship of the design criteria for lighting installations', presented at a UK Illuminating Engineering Society (IES) meeting in December 1967 (Reference 2.45). The author limits his attention to working environments with more than three fittings; he reviews accepted criteria for modelling, glare, scalar illumination, brightness patterns, and bounding surfaces, and considers what happens when the attempt is made to satisfy all these at once. He concludes 'that this is likely to happen during a large part of the working year in rooms with reasonably light-coloured decorations, some (but not too much) daylight from side windows, and lighting fittings somewhere in the range from classification BZ3 with 10% upward light to classification BZ6 with 50% upward light. For a room of a given size with given reflectances, the range of acceptable fittings will be much narrower, and limiting glare index will normally define a minimum proportion of upward light for a particular fitting . . ; if the reflectances—especially that of the floor cavity—are at the lower end of the acceptable range, there is scarcely any choice of fitting.' The whole of the discussion that followed the paper is worth reading in *IES Transactions*. Much of it is an agonised response to the next sentence in the paper's conclusion: 'It would therefore appear that, where all criteria apply, the technique of lighting may be capable of little further development, except in so far as new sources may make it possible to design smaller and neater fittings, and to obtain light of good colour-rendering more economically than at present.'

On the one hand, then, we have Dorsey's 'system', stimulating to consider but difficult to apply because it accepts the open-endedness of the situation, and, on the other, Jay's analysis, which (while admittedly limited to large and medium-sized working environments) leads not so much to a standard method as to a standard solution.

It is tempting perhaps to think that Jay's paper was a very determined exercise in *reductio ad absurdum*, but the author's sincerity in defending the basis of his sums is unquestionable. His critics 'should consider very carefully how an installation which does not comply with some of the criteria . . can be superior to one which does comply with all of them'.

The naïve comment might be that, if people like them better, they are superior. Impeccable experimental technique may establish the preferences of groups of observers presented in the laboratory with isolated aspects of visual conditions. But, if acting on preferences established in this way means that (with a little exaggeration) all lighting schemes must be the same, it becomes necessary to weigh their value against the desire for variety, which seems a spontaneous human reaction.

So we return to that last balloon: 'relate, evaluate'. 'Only connect', perhaps.

If we are content to regard a situation as presenting a routine problem, a routine solution is appropriate. At the 1972 UK national lighting conference at Warwick, there was some discussion—stimulated by a comment in J. M. Waldram's opening paper—on the value of the 'pattern book' approach. A distinction made later in the conference between 'lighting providers' and 'lighting designers' may apply here. The provider finds the pattern book useful, but the designer insists that every problem, at least in some aspects, is unique, and something better than the routine solution may be found by thinking of it in this way.

It is probably superfluous for me to declare, at this stage of the book, that this is my approach. I find it difficult to conceive of a situation that is so completely routine as to be incapable of improvement following some consideration of its particular features. I accept, of course, that the value of that improvement may not justify the cost of individual treatment, so that the need for pattern-book solutions—or for readily applied routine design methods—may be a fact of economic life. This is probably a matter of personality, and we are approaching the difficult question of whether 'the good' is an enemy of 'the best'. I am aware, too, as a practising designer of lighting schemes, that I may seem to be inflating the importance of my own activity—on the defensive, perhaps, against takeover by computer.

Nevertheless, this book is based on the belief that every lighting design problem has a unique context, but no unique solution. This is why the detailed reporting of completed, particular lighting schemes—such as appears in *International Lighting Review*—seems to me of greater value than the production of recipes for hypothetical and generalised situations. Such reporting tries to describe the context, to give at least an indication of special factors affecting the brief, or a clue to elements in the background, possibly unformulated or unspoken, which influenced design decisions. It does not say 'this is how you light a factory' but rather 'this is how a particular factory, with a unique context, has been lighted'.

Of course, we need to identify principles, to understand some of the reasons for success and failure. There is, I hope, a place for a book that is not simply an album of collected installation reports. I did not choose as a title 'How to design lighting schemes', but rather 'Lighting design in buildings', and the difference is more than a quibble: not 'what you should do is . .' but 'some of the things you might bear in mind are . .' And the balance between those things, the weighting they are given, in their unique context, is beyond analysis. It can be resolved only by human experience and judgment.

On a possibly less pretentious level, devices such as check lists clearly have a value, if only in confirming that some of the relevant factors have not been overlooked. The 1973 edition of the IES Code adopts this approach. Even simpler, the headings of the chapters in Part 1 of this book could be used in this way. When a sketch design has been developed, the check is applied: have we given due consideration to illumination, glare, brightness balance, colour, modelling, variety, flexibility, cost and maintenance? But what is 'due consideration'? Back again to that last balloon, 'relate, evaluate'. In other words: apply human judgment to the particular case.

In Part 4 of this book, beginning in the next chapter, we look at particular building types. There is certainly no attempt to summarise all you need to know about lighting factories in Chapter 23, or to give a recipe for the effective treatment of churches in Chapter 26. We shall see how some of the principles discussed earlier find application in these more limited contexts; we shall look at some examples of what has been done; and we shall find that every attempt to pursue a design idea raises questions about the particular situation.

Part 4

PARTICULAR BUILDING TYPES

Chapter 20

Civic and public service buildings

This last part of the book has even less claim to being comprehensive than the earlier sections. We cannot hope to treat fully the lighting of civic buildings, for instance, in a couple of pages. But what we can do is to look at some of the problems—even if little more than a random selection— that arise in the context of a particular building type.

Government and local-authority offices are, in functional terms, quite closely related to commercial office buildings. The image they present to the public has different priorities, but needs today to be considered with just as much care. The taxpayer and ratepayer will be impatient equally with an outmoded picture that suggests inefficiency and with apparent extravagance. The rather stuffy civic dignity that once characterised townhalls and the like is rightly disappearing. Outgoing, participatory aims encourage the citizen to feel involved, not merely as a suppliant or revenue provider, but also, if only indirectly, as the management. This is his building, and he wants to see what is going on in it.

The visual openness that follows brings with it the need for visual discipline, and the effect of the interior lighting as seen from outside should be emphasised in any check list applied to civic buildings. Continuity in the pattern of the layout around one floor, and from floor to floor, is important; and, where it is broken, it should help the onlooker to understand the functions of the various parts of the building. A townhall that is easy to read—reception, general offices, council suite and so on—contributes to civic awareness. Even so, bringing the building together is likely to be the preoccupation at the design stage. The use of only one tube colour is indicated, and variety reduction generally can help; tungsten fittings might be confined to specific areas, such as those with a stronger social content. At Hemel Hempstead townhall, recessed cold-cathode tubing, with independent control, runs right round the perimeter.

The specifically social building or complex provided by the local authority also raises the question of balance. The scale can vary considerably, the place can be called a community centre, an amenity centre, or be given a name. It may include sports facilities, a library, a theatre—but it should appear neither a municipal memorial nor a commercial fun palace. This calls for restraint in exterior lighting, in poster display and so on, but also for a determination to avoid the institutional inside the building, with the interiors seeming varied and inviting.

A superficial reading of the last two paragraphs might suggest that what I am saying is that, in a townhall, a very limited number of types of fitting should be used to give the building unity, but that, in another kind of civic building, lighting should add to the variety of the spaces. Clearly, no such generalisation is possible, but it does bring out the point that one of the fundamental issues to be settled in preliminary discussions between architect and lighting designer is this matter of unity and variety. One's only general comment is that, in practice, very few buildings err on the side of austerity, and a great many would be improved visually by a reduction in the number of different lighting-fitting types.

The civic building most likely to cause despair in those responsible for its design is the multipurpose hall. To the councillors on the committee charged with producing the brief, the idea of a high utilisation for a large interior that can be adapted for a number of activities has an obvious appeal. The danger is that the needs of none of the activities are really adequately met. Nevertheless, mechanical and electrical flexibility in the lighting can help in setting a different scene—this has been discussed to some extent in Chapter 8. What lighting cannot do is to make an interior with a flat floor suitable for an activity needing a raked one, and vice versa.

The courtroom is a public-building interior which makes special demands. Judge and jury must be able to observe nuances on the face of the accused and the witnesses. Counsel need all the help they can get in following the judge's reactions, and each other's. People must be able to read notes and records easily, and to examine exhibits. At the same time, there is a growing feeling that distraction from outside must be avoided; while rooflighting, typically through a dome, is traditional, pressures of space are producing basement or otherwise totally enclosed courtrooms, inwardlooking with the sense of confinement increased by necessary security. Yet it is in daylong sessions under these conditions that human beings are expected to reach conclusions of detached understanding and fairness. The professionals spend much of their working lives here, while the amateurs present are unsettled and perhaps alarmed by the occasion. The general effect of any lighting scheme must be bright, open and cool (400 lux as a minimum, any dark finishes limited in extent, de luxe tubes of 4000 K colour temperature). Walls should have as high a luminance as is consistent with avoiding any suggestion of a silhouette effect. Special attention must be given to the modelling of faces on the bench, in the witness box, and in the dock (particularly from the jury's viewpoint), but the necessary relief must be achieved without subjecting any party to discomfort glare. Where there is no daylight, some ability to change the visual conditions seems essential, whether this is by a very slow continuous modulation or simply by different switching after the lunch adjournment.

In buildings for the police or for fire services, and in those for water boards and other public utilities, a sense of being at the ready is important, both for efficient operation and for public confidence. Many areas therefore need lighting that is permanently in circuit; long-life lamps and low-maintenance fittings quickly repay any initial premium. Operations or control rooms where personnel have to scrutinise t.v. screens, luminous dials, and the like, need related brightnesses in the surroundings; this applies too in telephone exchanges with luminous indicators.

The solidity of many public buildings makes them liable for remodelling or refitting, rather than replacement. Lighting can prove effective in cutting the pomp and circumstance, and in creating a less formal, more relaxed impression.

Chapter 21

Commercial and office buildings

Beyond the necessary provision of basic services, display lighting in shops is more the concern of the user of the premises than of the designer of the building. Nevertheless, some understanding of display techniques is valuable to anyone concerned with lighting, since it is in display that many of the important but intangible elements in lighting design appear most clearly. There is some discussion of this in Chapter 27, in relation to art galleries and museums; but, for the moment, let us leave this aspect of shoplighting simply with references to accounts elsewhere (2.46; 2.47; 2.48; 2.49).

Paradoxically, whereas the small shop may appear to the building architect merely as a gap to be filled by others, he is likely to be more involved with detail and finishes in a large store. One basic decision will be over daylight. Normally today this is excluded from sales areas, but the perimeter can be used as a buffer zone between inside and outside, taking offices, fitting rooms, stores and so on. Although natural light is probably welcome in these areas, there is a contradiction in giving windows prominence in the façades of what appears as a blind building to most people inside it. A false façade of vertical metal elements or some other abstract treatment is appropriate, and can well serve a thermal purpose too. A large store is a suitable building for an integrated system of lighting and air conditioning, though lamp heat can sometimes be used more directly, as with warm air ducted straight from display windows. Sweeping claims are made from time to time about the control of customer movement by lighting; it is true that brightness has a basic attraction, but glare tends to repel, and it might be unwise to build too much on the shaky foundation which is all we have so far.

The typical commercial building is the office block, in many ways the building type most characteristic of our time. Much of the earlier discussion in this book on criteria and design has thus tended to be in terms of office buildings. One or two related ideas are pursued later in this chapter, but, first we can usefully take a conventional general office as an example of the lumen method of computation (further examples can be found in Chapter 23, in illuminating-engineering textbooks, and in 'Interior lighting design' (Reference 1.2).

Reducing the situation to barest essentials, let us assume that we have an office 8·4 m long, 4·8 m wide and 3·25 m high. An illumination of 400 lux is to be provided using enclosed plastics diffuser fittings for two 1·8 m 85 W Natural tubes, surface-mounted. The working plane is conventionally taken as 850 mm above the floor, so that the situation is as in Fig. 4.1.

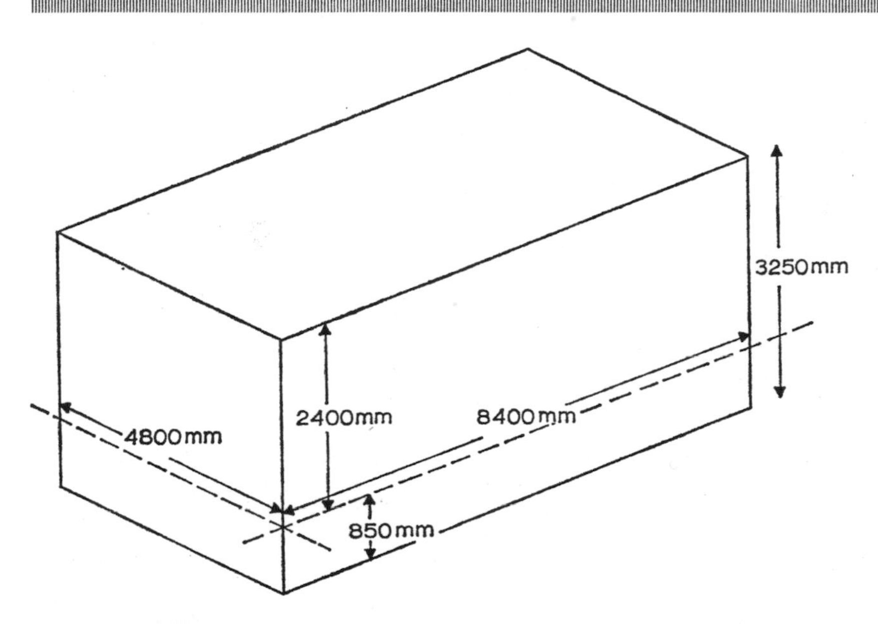

Fig. 4.1 Key dimensions in general office scheme

The room proportions are expressed as a room index k_r, which is defined in terms of the length and width of the room and the mounting height H_m of the fittings above the working plane.

Here

$$k_r = \frac{L \times W}{H_m(L+W)}$$
$$= \frac{8{\cdot}4 \times 4{\cdot}8}{2{\cdot}4(13{\cdot}2)}$$
$$= 1{\cdot}27$$

Additional data that we need to know are the reflectances of the main surfaces and the basic performance information on the fitting as described in Chapter 12. If the reflectances here are

$$\rho_c = 0{\cdot}7$$
$$\rho_w = 0{\cdot}3$$
$$\rho_f = 0{\cdot}15$$

fitting data:		
	u.l.o.r.	20%
	d.l.o.r.	50%
	BZ number	6
	luminous area	2700 cm^2

The principles on which the lumen method rests have been discussed in Chapter 10. The central relationship is

$$\text{flux installed} = \frac{\text{flux received}}{\text{utilisation factor} \times \text{maintenance factor}}$$

A table published by the manufacturer of the fitting being used tells us that, for a room index of 1·27 and reflectances $\rho_c = 0{\cdot}7$ and $\rho_w = 0{\cdot}3$, the utilisation factor is 0·42. The flux received is the illumination (the lumens arriving on one square metre) multiplied by the area. The maintenance factor is a conventional 0·8 (Chapter 10).

Thus

$$\text{flux installed} = \frac{400 \times 4{\cdot}8 \times 8{\cdot}4}{0{\cdot}42 \times 0{\cdot}8}$$

$$= 48\,000\,\text{lm}$$

The 'lighting-design lumens' of one 1·8 m 85 W Natural tube are 4150, so that the installed flux represented by each twin-tube fitting is 8300 lm.

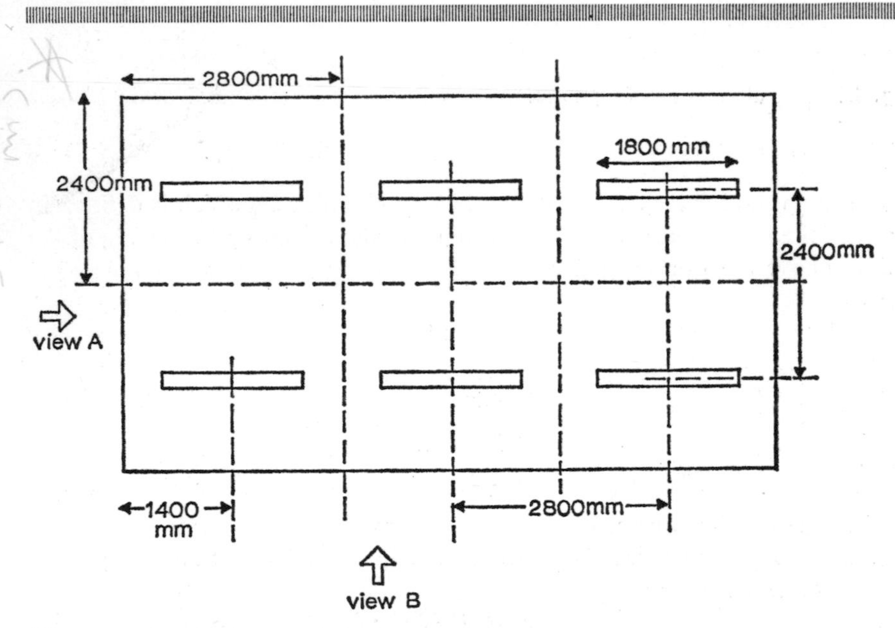

Fig. 4.2 Regular layout of fluorescent fittings

49

Plates 49 & 50 Fittings layout follows furniture in council chamber and board room

50

52

Plates 51 & 52 Banks in Malta and Japan, both with rectilinear fittings closely related to fabric of interior

Plate 53 Court room without daylight. Flooding of feature wall prevents 'directionless' feeling that could result from use of diffusing ceiling in almost square interior

Plate 54 Despite extensive glazing, working illumination in most recent office buildings is provided essentially by electric sources

Thus

$$\text{number of fittings required} = \frac{48000}{8300}$$

$$= 5{\cdot}8 \text{ or, in practice, } 6$$

A possible layout seems to be two rows of three fittings, as in Fig. 4.2.

The spacing between the centre of a fitting and the wall is half that between centres in a conventional layout of this kind. The fitting-to-fitting spacing is here 2800 in one sense and 2400 in the other. Since the limiting spacing for a BZ6 fitting is 1·5 times the mounting height (Chapter 7) this layout is acceptable in uniformity terms.

The scheme remains tentative, however, until the glare index has been checked and compared with the limiting glare index quoted in the IES Code—here, for general offices, 19. The computation is merely outlined here; it is set out in detail in 'Interior lighting design' (Reference 1.2) and elsewhere.

The height of the fittings above a 1.2m eye level can be taken as 2m. To find glare indexes, we have to express the plan dimensions of the room in terms of this height H. Thus, for view A, the dimensions X, across the line of sight, and Y, along or parallel to that line, are given by:

$$X = \frac{4{\cdot}8}{2} = 2{\cdot}4$$

$$Y = \frac{8{\cdot}4}{2} = 4{\cdot}2$$

A table in, for instance, 'Interior lighting design' tells us that in these circumstances (which include the 'dark floor', $\rho_f = 0{\cdot}15$) for an endways view of a BZ6 fitting, the 'universal initial glare index' is 15·7. Subsequent graphical data show the corrections to be applied to this figure for differences from the situation for which the original table was computed in room reflectances, flux-fraction ratio, downward flux, luminous area, and mounting height. Collecting these corrections together,

	+	−
Reflectances and f.f.r. (70%, 30%, 0·4)	3·6	
Downward flux (0·5 × 8300 = 4150lm)	3·7	
Luminous area (2700cm^2)		5·0
Mounting height H (2m)		0·5
	7·3	5·5

The total correction to be applied to 15·7 is thus +1·8; so the resultant glare index is 17·5—less than the limiting value of 19 given in the code.

For the view B, X is now 4·2, Y is 2·4, and the view of the fittings crossways. Proceeding in the same way reveals a glare index of 17·0. These results, or figures close to them, might alternatively be found using one of the slide-rule calculators.

The conclusion is that the scheme is acceptable in terms of illumination and direct glare, the two criteria most widely considered—or, at least, the two that are subject to numerical checks in orthodox illuminating engineering.

It is worth noticing, in passing, if the linear fittings here had been aligned the other way (parallel to the shorter side of the room), the glare index for view A would have been 19·5—above the limit. This does not, however, mean that fluorescent fittings are always less glaring seen end on; it depends on their light distribution. Except in very small rooms, a metal trough reflector classified in BZ4 is better from the side, and for a plastics reflector there is often little to choose between the two viewpoints.

As a further numerical example of simple office lighting, let us take a scheme using recessed modular equipment. Consider a room 7·8m by 6·6m, with a suspended ceiling 400mm below the structural ceiling height of 3250mm (Fig. 4.3).

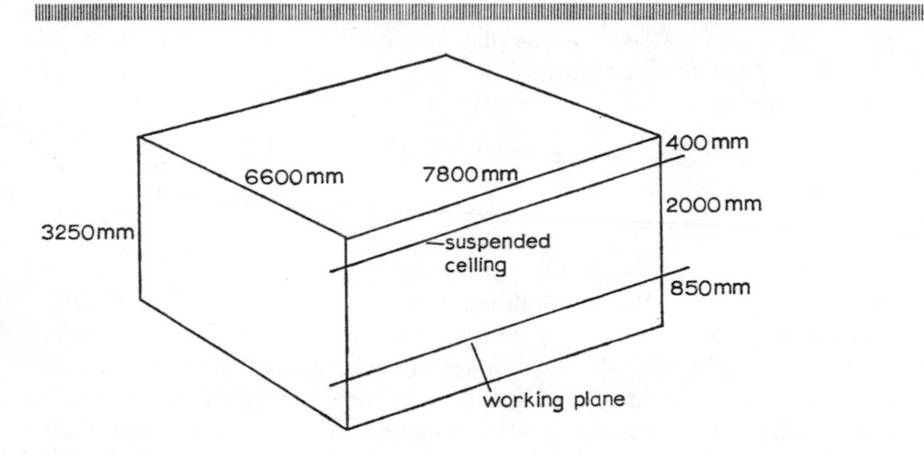

Fig. 4.3 Office with suspended ceiling

If the ceiling is made up of 600mm square tiles, there will be 13 rows of 11, and the lighting equipment must fit into this grid. The units will replace one, two, three or four tiles depending on the tube lengths for which they are designed (nominally 600mm; 1200mm; 1500mm or 1800mm; 2400mm). There would have to be a special reason for using the smallest of these, as the large number of fittings and lamps is inevitably uneconomical. We shall look at the other possible sizes.

Assume that the illumination is to be in the 500–600lux range; that the tube colour is again Natural (a reasonable standard for offices generally); and that the reflectances are 70% for the ceiling, and 30% effectively for the walls, with a 'dark' floor.

Purely geometrically, there is a large number of possible layouts. However, the smallest acceptable number of fittings under a given set of conditions is likely to imply least cost. The smallest number is normally associated with the maximum spacing. The guidance given earlier (Chapter 7) may be repeated:

	Maximum spacing/mounting-height ratio
BZ1 and BZ2	1 : 1
BZ3 and BZ4	1·25 : 1
BZ5 to BZ10	1·5 : 1

It is unlikely that a fitting classified as BZ1 or BZ2 would be needed here, unless a very low glare index were sought for some reason. Recessed fittings with louvers or prismatic plates are often BZ3 or BZ4, while, with opal diffusing dishes, BZ5 is usual. For our present purposes, we may conclude as follows, for the mounting height of 2 m above the working plane:

	Maximum spacing between centres	
Light control medium	Fitting to fitting	Fitting to wall
	m	m
Louvers or prismatic plate	2·500	1·250
Opal diffuser	3·000	1·500

These are rules of thumb, and they need qualification for long fittings at relatively low mounting heights, as we shall see. For convenience, let us restrict our attention to the six layouts shown in Fig. 4.4.

In *a*, there are two rows of three 2·4 m fittings, necessitating cutting of some tiles within the rows. However, the layout can be dismissed at once as the spacing between rows, 3600 mm, is excessive. If the six fittings are turned through a right angle to produce *b*, the spacing is just acceptable for fittings with opal diffusers; although, according to the rule of thumb, the distance of 1·8 m from centre to long wall is too great, the fact that the source extends to 600 mm from the wall is what matters. In such circumstances, we can accept a distance of one half of the length of the fitting as a gap between BZ5–BZ10 types, and one-third for more concentrated distributions from louvers or prismatic plates.

For most tube sizes, the overall length of the fitting exceeds the nominal multiple of 600 mm; this means that recessed modular types have an extended lamp housing above the ceiling plane, and so continuous mounting is not possible—hence arrangements such as that in *a*. However, 1800 mm fittings can take both 1500 mm (65 W) and 1800 mm (85 W) tubes with lampholders etc., within the length of three 600 mm tiles. Continuous mountings, as in *c*, becomes possible. Here the spacing is all right for BZ5 or higher, but not consistent with conventional uniformity for more concentrated distributions. The layout in *d*, also of nine fittings, is, however, acceptable for all classes with a BZ number of 3 or more.

Similar conclusions apply to layouts of 1200 mm fittings. Scheme *e* is acceptable with opal diffusers, but the distance from the centre line of the row to the adjacent endwall is excessive for louvers or prisms. Scheme *f* passes the more demanding test.

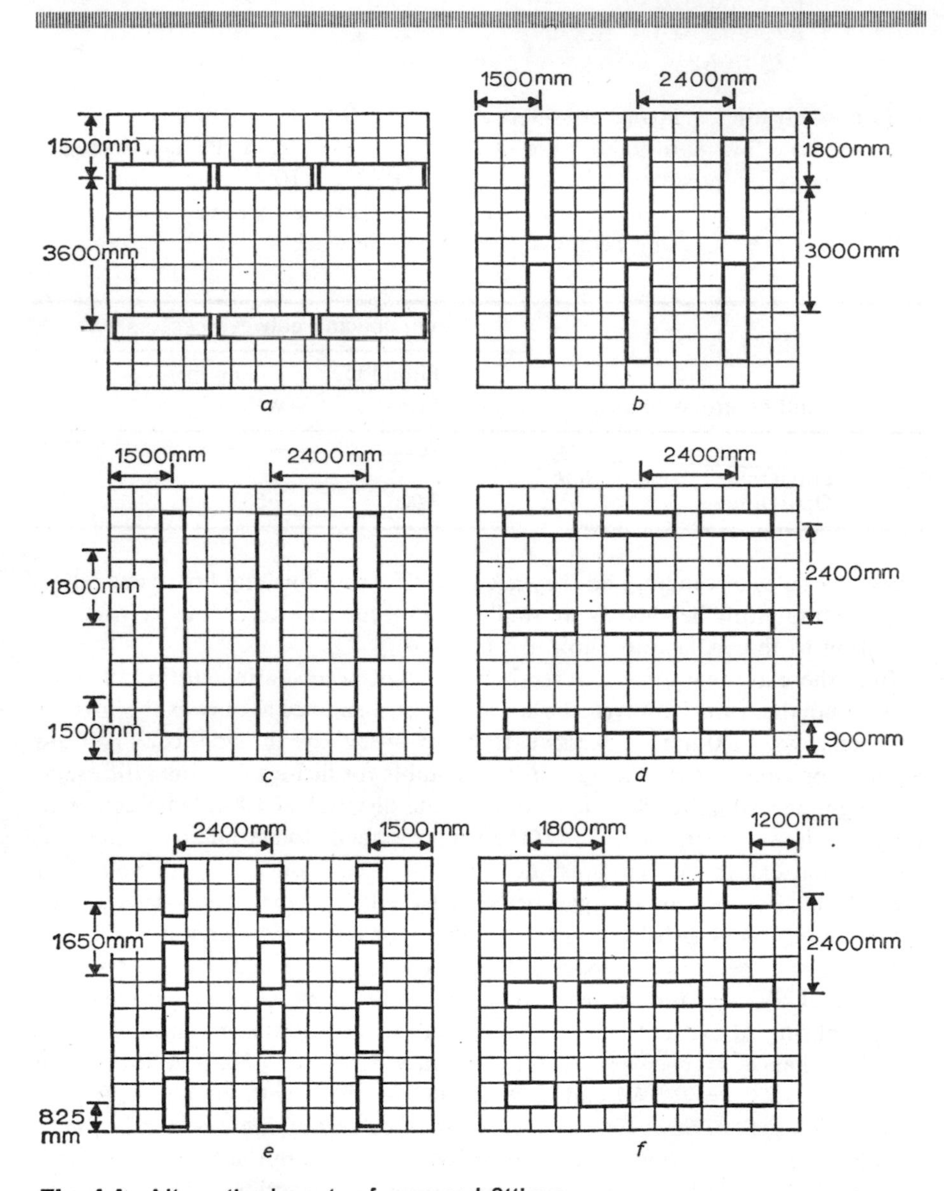

Fig. 4.4 Alternative layouts of recessed fittings

These results may be summarised in a table:

Scheme	Acceptability	
	Opal diffuser	Louvers and prisms
a	no	no
b	yes	no
c	yes	no
d	yes	yes
e	yes	no
f	yes	yes

So far in this discussion, we have not considered the number of tubes needed in each fitting to produce an illumination in the specified range (500–600lux), and we should look at this now to see whether all the schemes we have considered are feasible from this point of view.

The utilisation factor will vary with the light control medium, and, to some extent, with tube length and the number of tubes per unit. But the figure for most fittings of this type with the room proportions and reflectances here is close to 0·4. If the maintenance factor is about 0·8, we can take the product of the two factors as one-third; the installed flux must therefore be three times the flux received. Thus, with a design level of 550lux,

$$\text{installed flux} = 3 \times 550 \times 7{\cdot}8 \times 6{\cdot}6$$
$$= 85000\,\text{lm}$$

Dividing this by the output of the Natural lamp for each possible rating, we find the total number of tubes required:

Rating	Nominal length	Lighting-design output	Rounded-off number of tubes
W	mm	lm	
125	2400	6100	14
85	2400	4800	18
85	1800	4200	20
65	1500	3300	26
40	1200	2000	43

For the scheme *b*, with six 2400mm fittings, the 125W rating is not convenient, as the choice lies between 12 tubes and 18 (which would give, respectively, 470 and 700lux); but six fittings for three 2400mm 85W lamps are appropriate. In schemes *c* and *d*, both with nine fittings, the choice is between two 85W and three 65W tubes in each; the former gives about 495lux, probably close enough to the specified level, and the latter 570lux. In the schemes for 1200mm lamps, four per fitting produces 620lux and three per fitting 460lux.

The choice between alternatives is influenced by factors peculiar to the particular case. One lamp rating could be favoured because its exclusive use elsewhere in the building had already been decided on. The rhythm of the windows could make four fittings in the length of the room preferable to three, or vice versa; but, other things being equal (as they never are), the alternatives could be reduced to three:

Scheme	Lamps	Total load, including gear	Illumination
		W	lux
*b*2	6×3×85W(2400mm)	1800	550
*d*1	9×2×85W(1800mm)	1800	495
*d*2	9×3×65W	2160	570

However, since scheme *b* is acceptable in terms of uniformity only with opal diffusing panels, its glare index should be examined. The computation yields a figure of 19·5, exceeding the limit of 19. While BZ4 and BZ3 fittings would give indexes of about 17 and 16, respectively, the spacing limitation rules these out here (see table of acceptability above). The choice therefore narrows to the alternative lampings of scheme *d*, which means 495lux from 1·8kW or 570lux from 2·2kW. The former is marginally better value in lux per watt, and since it involves 18 lamps rather than 27, it is preferable in installation and maintenance terms. So the final choice looks like scheme *d*1: nine 1800mm fittings, each for two 85W tubes. It may be worth noting, however, that, had the floor been light (30% reflectance), scheme *b*2 would have given a glare index of 17·9, well within the limit. The irony is that, in use, with much of the floor covered with desks and much of those desks covered with paper, the effective reflectance of the 'floor' might be much higher than that in the empty room which is all that traditional lighting engineering tends to consider.

A related point can readily be observed in practice. In a large office building using a constant system of recessed fittings throughout its open-plan working areas, the ceilings in drawing-office zones can be noticeably brighter than those elsewhere, because of the increased reflection from the boards. This has the happy result that a standard lighting scheme has effectively a lower glare index in those areas where the IES Code makes such a recommendation.

The effect on lighting design of special glasses is one that can appropriately be introduced under the 'offices' heading. Whereas much current discussion on low-transmission glazing is concerned mainly with thermal effects, there is always an associated reduction in visual intensities, and usually one which appears differentially across the spectrum. What needs emphasis is that these visual effects are not merely unavoidable consequences of thermal decisions, but phenomena with positive value in lighting design, particularly in terms of brightness balance.

We have seen earlier (Chapter 14) that when electric lighting for use in the presence of daylight was first investigated systematically, model studies suggested that levels of 400–600lux were appropriate for office rooms with a depth of

7–10 metres. The criterion was a satisfactory balance with the brightness of the sky and the external scene. An empirical relationship was developed for the desirable supplementary illumination in terms of the average sky luminance and the daylight factor (IES Technical Report on daytime lighting, Reference 2.4). In SI units,

$$E = \frac{\pi L D}{10}$$

where: E = illumination, lux
L = average sky luminance as seen through window, cd/m^2
D = daylight factor, %

If the 'standard overcast sky' condition, often assumed in numerical work on daylighting, is taken,

$$\pi L = 5000$$

and so

$$E = 500D$$

A room in which the daylight factor falls to 1% needs supplementary illumination, on this basis, of 500 lux; if the daylight factor at the back is 0·7%, the supplement is 350 lux. It may seem surprising that, with less daylight, a smaller supplement is required, but this emphasises again that p.s.a.l.i. recommendations are primarily guided by visual balance considerations—and that the objective in p.s.a.l.i. design is to maintain the general impression of a daylit space. It is clear that this becomes impossible with daylight factors less than, say, 0·5%. For this figure, the supplement would be 250 lux and a significant area at the back of the room would have a total level below the 400 lux recommended as a minimum for office work; this would itself rule out the p.s.a.l.i. approach in these circumstances. At some point, then, depending on the activity, 'permanent artificial lighting', or p.a.l., becomes necessary—and certainly with offices having working positions 10 m or more from the window wall. The brightness caused by daylight remains a kind of independent standard—whether it is of the external landscape, of the internal area near the window, or of the sky itself, which may be all that you can see from points well back into the room. Balance with this natural brightness remains a key factor in settling the illumination to be provided, in the immediate human terms of preventing an impression of deprivation and gloom in the parts of the room well away from the external wall. It is practice rather than any theoretical study that has led to 1000 lux being regarded as some sort of magic figure, and the round number is part of the appeal. What is probable is that the illumination needed for a happy relationship with the exterior brightness is greater than what would otherwise be provided in terms of current standards for performance or preference. In other words, it is the presence of the sky brightness, or some direct consequence, that is leading to the need for 800–1200 lux, depending on the detail of window design, surface finish, and so on—levels which exceed conventional present-day provision on grounds of efficiency or amenity. The alternatives seem to be those of accepting this situation, if conventional windows are used, or taking steps to limit the effect of daylight by reducing the sky brightness as it appears from inside the building. Conventional blinds or curtains interfere with the view, one of the main functions of windows in this context. Sheers or other fabrics which allow some vision, represent a possible solution, particularly where they are appropriate to the furnishing style of the interior; devices such as louver blinds have the virtue of adjustability. But the most positive way of controlling apparent sky brightness—and the most committing—is by the use of reduced-transmission glass.

Details of the glasses available are published by the manufacturers. There are three main categories. The older type of heat-absorbing glass transmits in the 6mm thickness a total of about 60% of the solar energy. The green version admits a higher percentage of visible light, the grey a lower, and the absorption occurs in the body of the glass. More recent developments include a bronze surface-tinted type, admitting about two-thirds of the solar energy and half the light, and various laminated types with more restricted transmission. For comparison, 6mm clear glass permits about 85% of the light to pass through, and a similar percentage of the total solar energy.

Thus, if the brightness of a particular unobstructed area of the sky is, say, 2000cd/m^2, its luminance seen through conventional single glazing is:

$$2000 \times 0{\cdot}85 = 1700\,\text{cd/m}^2$$

Through double glazing with two leaves of 6mm clear glass, this becomes

$$2000 \times 0{\cdot}85 \times 0{\cdot}85 = 1400\,\text{cd/m}^2$$

(It is unrealistic to express the result to more than two significant figures). In the same way, we find that, through surface-tinted glass, the observed luminance will be about 1000cd/m^2, and, through double glazing consisting of one leaf of clear glass and one surface-tinted, 850cd/m^2.

If 1000lux from electric sources is needed in a relatively deep office with single clear glazing, half this figure should be adequate in terms of brightness balance for the last combination—which is quite likely in practice. Or, since the sky luminance will sometimes be higher, a level in the 600–700lux range is a reasonable recommendation. With de luxe tubes, this implies a loading of 35–40W/m^2, which is manageable, and indeed useful, thermally. In summary, then, the purpose of the double glazing is to reduce conduction loss and condensation, and that of the special glass is also two fold: reduction of gain from solar radiation, and control of apparent sky brightness.

Subjective reactions from the users of the building to the visual effect of low-transmission glasses have not been studied systematically; existing evidence is on an anecdotal level. One of the few negative reactions I have encountered was in a building where the original conventional clear glass had resulted in unacceptable discomfort in summer. The windows were reglazed with a reflective laminate type, and, though thermal conditions were undeniably better, the office occupants complained of a 'dreary and depressing' effect. This case,

55

56

Plates 55, 56 & 57
Surface-mounted fittings in offices. Stock Exchange floor is lit by black-trimmed fittings for four 40 W tubes above opal diffusers. Open-plan office area has twin 65 W fittings with prismatic light control. In drawing office, lines of fluorescent fittings are broken by mercury-lamp types for variety of appearance and effect

57

58

Plates 58 & 59 Lighting integrated with ceilings of strong visual interest: a bank in Birmingham and offices in New Delhi

59

Plate 60 Lighting integrated with staircase in London office building

61

62

Plates 61 & 62 Suspended and recessed sources in commercial reception areas. In higher space, cylindrical housings for tungsten lamps provide general illumination, and a continuous baffle along relief wall screens fluorescent tubes and low-voltage spots. An alternative approach, where ceiling is lower, has fluorescent lamps above low-brightness specular louvers for general light, projecting clear acrylic cylinders below tungsten lamps within ceiling space for decoration, and a series of sequentially switched PAR38 reflector lamps, some coloured, above circular apertures in front of pictorial display however, is exceptional in many ways—a 'first aid' treatment, knowledge of the previous conditions, openable windows permitting direct visual comparison, all combined with a glass of markedly reduced transmission. In buildings where reflective laminate glass has been an essential feature of the design, there does not seem to have been any comparable reaction, and the general experience with bronze surface-tinted glass in sealed windows is that many of the building users have not been conscious that the glazing was special in any way.

Reduced-transmission glasses naturally affect the external appearance of the building. Even quite a modest reduction tends to unify the façade and to conceal what might otherwise seem an untidy interior. The laminate varieties have a pronounced mirror effect—people working at street level have reported passers-by approaching their windows apparently to stare in but, in fact, to check their own appearance.

Chapter 22

Transport, traffic and storage buildings

Any lighting related to transport or traffic must recognise the fact that luminous signs and signals play a vital part in its safe and efficient operation. Not interfering with or confusing the message such media carry must be a major preoccupation in design. Admittedly, this applies with greater emphasis to exterior lighting, but, in railway stations, bus depots, multistorey car parks, garages and the like, general illumination should be developed in terms of controlled brightnesses that permit immediate appreciation of the significance of signs and signals. Illumination recommendations may be no more than indirectly useful. If 50lux is proposed for interior parking areas, we can interpret this as meaning something like 3W per square metre of fluorescent lighting; so for an area of, say, 30m by 15m, we can expect a lighting load of 1300–1400W. This would mean ten fittings for twin 1·5m 65W tubes, and the design rests on the disposition of these fittings to reveal hazards and give necessary visual guidance more than on their regular arrangement to produce uniformity on some mythical 'working plane'. Here it is, in fact, unlikely that ten fittings would provide enough lighting points, and we could be better advised to think in terms perhaps of 30–35 fittings for twin 20W tubes. The fittings would preferably have sealed lamp enclosures in diffusing plastics (of an antivandal material such as polycarbonate if there is public access) and metalwork treated to resist corrosion. Particular local authorities may have more demanding requirements. In this whole field, it pays to make pessimistic assumptions about maintenance and about the physical conditions in which the equipment will be expected to operate, and go on operating over the years.

The change from steam to diesel has transformed railway termini and other major stations by removing the necessity for a large volume of air above the platforms. In the older form (e.g. Paddington, St. Pancras in London), the platforms, concourses, and other open areas are all part of one high, glazed-roof interior. In contrast, at Euston in London, or at New Street, Birmingham, the platforms are 'below stairs', machines for getting on the trains from, and all the other activity of the station centres round a detached main concourse, visually a high, impressive interior that might almost as well be associated with aeroplanes or ships. The glass-roofed shed is half outside; it is appropriate to light it from powerful high-mounted sources that provide a basic general level to which local supplements are added. The downstairs platform, on the other hand, has limited height available, and so demands more modest outputs in greater number—diffusing fluorescent fittings mounted parallel to the line, perhaps on continuous trunking. Up above, the main concourse, though still large, has a more protected feeling that goes with a more luxurious impression. Equipment with industrial applications is out of place. The height makes servicing from above welcome, and the concentration of a thousand watts or more in one relatively small lamp may make incandescent sources attractive in spite of low luminous efficacies (tungsten–halogen types are appropriate).

One way of avoiding servicing difficulties at considerable mounting heights is to provide the overall illumination indirectly, by projecting light up on to the ceiling. This demands a finish that will retain a high reflectance, but it has added point if there is strong structural interest. A system such as this, which has inherently low efficiency, needs powerful high-efficacy sources to make it economical. The newer types of discharge source naturally spring to mind, and in particular the question of the suitability of high-pressure sodium lamps arises. Sixty-four of the 400 W rating were, for instance, used in this way for general illumination of the main hall of Grenoble station, France, rebuilt at the time of the winter Olympics in 1968 (the installation is described in the *International Lighting Review*, Reference 2.50; the *Review* is the major source of information on completed lighting installations throughout the world, and is strong in the field of traffic and transport). Upward lighting of this kind produces a 'sky' of fairly uniform brightness, and so a soft effect. Local lighting at booking offices, kiosks, showcases, and so on, punctuates this with areas of emphasis and stronger modelling; but, in nature, the overall illumination, whether from blue sky or from clouds, is cool in quality, and the direct and specific effect is from sunbeams. If the upward light comes from high-pressure sodium, however, the general illumination is inevitably warmer in quality than is the local effect from tungsten or fluorescent—an inversion that may be recognised consciously only by the interested specialist but may leave the user of the space with a sense of unease. This is to some extent a personal reaction, but I should go further and argue that, whereas the warmth of high-pressure sodium is very welcome at the low levels of most exterior lighting, its interior use must be carefully considered if an oppressive effect is to be avoided. Certainly in the situation here—indirect lighting for a large station concourse—mercury would seem more appropriate. It may be possible to position the fittings on the tops of bookstalls, inquiry offices, local canopies at left-luggage counters, and so on. There is also a good case for switching to lower intensities after dark; the greater diversity and the increased emphasis the local lighting then achieves are somehow appropriate to the evening atmosphere, quite apart from there being no longer any need to balance the visible brightness from daylight.

Hangars, road-transport garages, and engine sheds, throw up the problems of any large industrial interior—access to equipment for maintenance, its suitability to the physical conditions, low reflectances, the difficulty of reconciling glare control with adequate vertical-plane illumination, and so on. Control rooms, particularly for air transport, have their own specific visual problems; a discussion of lighting in radar viewing rooms appeared in the January 1972 issue of *Light and Lighting* (Reference 2.51). It is true, of course, that a great deal of lighting for traffic and transport is designed by specialist engineers in the industry, and an architect concerned with related buildings needs a working liaison at an early stage. But some awareness of the problems and possible solutions contributes to one's general appreciation of lighting design—and thinking about the lighting encountered is one way of adding interest to a routine journey.

Lighting in storage buildings depends on the type of storage system employed. Again, there is little relevance in thinking of conventional working planes and uniformities. The important point is that the fittings and their disposition should result in light reaching the material in shelving, racks or bins, whatever the level of stock. To achieve this, the storekeeper is probably happy to accept some degree of glare as he walks about. Where there is continuous and varied activity, all the lighting will be needed throughout the working period. But where visits to the store are intermittent, a low, maintained illumination from some discharge source can be supplemented by tungsten when and where it is necessary. In cold stores, where the low ambient temperature is likely to reduce the output of fluorescent tubes, a more concentrated source in a heat-retaining enclosure is indicated; the psychological effect of apparent warmth could be welcome here. Automatic warehousing is being developed with varying degrees of sophistication. Where control is manual from a crane cab or some similar position, good general lighting of the whole space is clearly essential. But even with punched card or computer control, the need for high utilisation of expensive equipment makes quick identification of faults and attention to them urgent, and justifies the maintenance of high illumination—a point that applies generally to automated plant.

Chapter 23

Industrial buildings

As more machinery is used in offices, and as light industry employs increasing numbers of women in interiors that have received some attention in making them pleasant for work, so the old distinctions between office and factory surroundings are breaking down. With the coming of the large landscaped office, size is no longer a distinguishing feature; conversely, some extensive office spaces with regular lines of desks have been described, not unreasonably, as 'paper factories'. Some manufacturing processes, of course, make greater visual demands than does normal office work—the assembly of small electronic components is one example—and so justify illuminations in the highest ranges normally encountered in buildings. For economy, and to avoid having too great a heat load, this may mean a localised scheme, with perhaps 500lux generally from an overall spread of fluorescent fittings and an additional 1000–2000lux on the job from lines of reflectors related to benching and at a lower mounting height; but generally, lighting for offices and for light industry is becoming more closely related, with many of the same criteria applying.

Heavy industry, however, is a different story, and mainly because the available height is much greater. If fittings can be mounted 10m above the floor, they can be spread at this distance and more, and an interior measuring, say, 30m by 20m can be lit from only six points, with major economies in installation and maintenance. Let us pursue this as a further example of the lumen method.

Suppose 400lux is required and that the reflectances of ceiling and walls are 0·3 and 0·1.
Then

$$\text{room index} = \frac{LW}{H_m(L+W)}$$

$$= \frac{30 \times 20}{10 \times 50}$$

$$= 1{\cdot}2$$

Notice that the floor has been taken as the working plane here. For a particular type of high-bay aluminium industrial reflector, the utilisation factor in these conditions is 0·44.

We may take a maintenance factor of 0·75. Then

$$\text{installed flux} = \frac{\text{illumination} \times \text{area}}{\text{u.f.} \times \text{m.f.}}$$

$$= \frac{400 \times 30 \times 20}{0{\cdot}44 \times 0{\cdot}75}$$

$$= 720\,000\,\text{lm}$$

If there are to be just six points, the installed flux at each must be 120000lm; but the 'lighting-design output' of the 2000W colour-corrected mercury lamp is 118000lm, and the difference smaller than the likely errors in the other quantities. Such a scheme represents a lamp loading of 20W/m^2, or a total load, taking 80W as the loss in each ballast, of 12480W.

It is interesting to look at the alternative: tubular fluorescent lamps. For simplicity, assume that the same installed flux is required, namely 720000lm. On this scale, we should obviously consider using the tube with the highest output, and, of those normally employed, this means the 2·4m, 125W White lamp, lighting design lumens 8300. The largest conventional reflector fitting takes four of these tubes; so

$$\text{number of fittings} = \frac{720\,000}{4 \times 8300}$$

$$= 21{\cdot}6$$

Two nearly continuous rows of 11 fittings are indicated. The total electrical load, including control gear, is very close to that of the other scheme. The alternatives are shown in Fig. 4.5.

The mercury scheme *a* has just six points. The fittings cost £150–£200, and the initial lamping a little under £100. The 88 fluorescent tubes in *b* are in 22 fittings; corresponding costs are £600–£700 and around £55. Lamp life of the two types is about the same, so it would take a long while for the lower materials cost at lamp replacement to compensate for the higher initial equipment cost with the fluorescent scheme—but, in any case, the much higher labour costs for cleaning and relamping combine with the greater installation cost of labour and cable to make *b* undeniably more expensive. However, it can be argued that it is preferable, and management must decide whether the advantages are worth the extra cost.

What are those advantages? First, the colour rendering, even of high-efficacy tubes, is better. Secondly, the tubes strike after a negligible delay whereas the mercury lamps take some minutes to run up. Thirdly, any interruption in the electricity supply must be followed by a wait, again of some minutes, for the mercury lamps to cool down before they will restrike, and run up again (a supplementary system of other sources may be needed to bridge the gap).

Again, flicker may be noticeable from the mercury lamps. The main difference, however, is in light distribution. Because the high-pressure discharge system packs so many downward lumens into a relatively small apparent area of fitting, a lower BZ number is needed to meet any given glare-index criterion than is the case with the fluorescent fitting giving far fewer lumens from a larger area. The low BZ number implies a strongly vertical flow of light with possibly inadequate

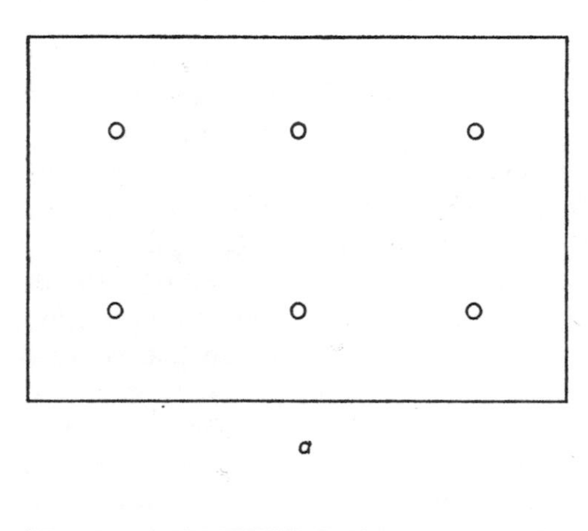

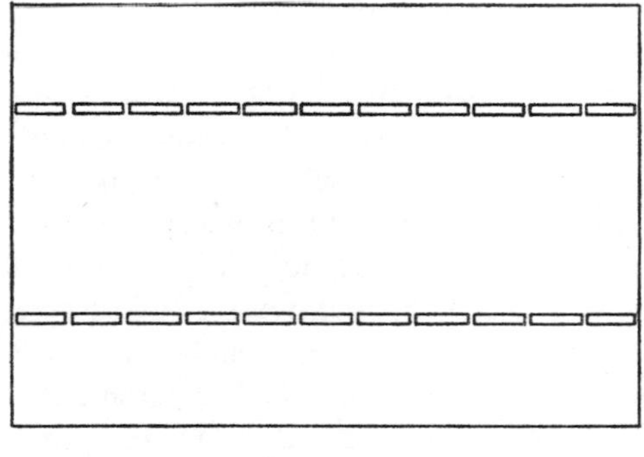

Fig. 4.5 Alternative schemes for high industrial interior

illumination on vertical surfaces. In other circumstances, high reflectances can soften this effect, but, in heavy industry, light finishes may not survive. Most detached observers presented with the alternatives would be likely to find the fluorescent scheme preferable subjectively.

In this example, a maintenance factor of 0·75 was assumed—a rather lower figure than would conventionally be taken for a typical commercial scheme, but it should be emphasised again that depreciation in output due to dirt inevitably depends on the cleaning interval as well as the situation. This point was mentioned in Chapter 10, where there is also a discussion of the importance of access to lighting equipment for maintenance, something that particularly applies in the industrial context. It is here, too, that we most commonly encounter hazardous atmospheres with the special demands they make. A useful summary of their implications for electrical equipment generally appears on pages 120–124 of the UK Electricity Council building-services handbook (Reference 2.52). It considers low flash-point areas, dust and grit, corrosive atmospheres, damp situations, high temperatures, mechanical damage, and fungus and mould growth. IES Technical Report 1 (Reference 2.1) covers lighting in corrosive, flammable and explosive situations.

Special lighting, whether for processing or inspection, is associated with a number of manufacturing industries. The techniques have been developed over the years, largely empirically, and are better regarded as part of production engineering than as building design. Nevertheless, the factory building must facilitate their application. Close liaison with the client's engineers over this becomes part of the designer's general concern with processing requirements.

Some processes require very precise control of physical conditions. In the Sangamo Weston factory at Felixstowe, England, for instance, where precision electrical instruments are made, the specified conditions in production areas were that the air temperature should remain within 2°F of 68°F, and the relative humidity 45–55%; in the 'white area' the relative-humidity tolerance is only ±2% (see Appendix H in Reference 2.35). The response to this kind of demand is to separate the internal environment as completely as possible from what happens outside, and the 'closed box' windowless factory results.

The arguments in favour of daylight or no-daylight in single-storey factories have been rehearsed in Chapter 14, but some further aspects of the windowless factory should be mentioned here. It is only relatively sophisticated managements who consider this approach, and their enlightened self interest is likely to result in environmental provision exceeding the typical; so it is difficult to know whether the good standards of amenity characteristic of no-daylight factories are in response to a feeling that some compensation for deprivation is desirable, or whether they would have been provided by these managements in any case. The appendix mentioned above to the Greater London Council research paper on windowless environments (Reference 2.35) includes some answers to a questionnaire circulated to managements and architects. Part of question 9 reads

> '(*a*) What has been the reaction from persons at all levels in the windowless environment?
>
> '(*b*) Has there been any effect on the levels of absenteeism, sickness, quality of work, performance . . .'

and question 10

> 'Has any special consideration been given to the problems of the internal visual environment and the standards of internal architecture?'

The response from Sangamo Weston Ltd., about their Felixstowe factory, includes

Plate 63 Perimeter lighting increases apparent size of auditorium in one of new generation of smaller cinemas

Plate 64 As with top of stage setting, upper part of theatre auditorium is sometimes intended to disappear into darkness with no defined boundary

Plate 69 In social or entertainment context, uniformity may be consciously discarded in favour of liveliness that is created by people moving in and out of patches of light

‘9 (*a*) On the whole, extremely favourable. It is a fact that one is quite unaware of the absence of natural daylight, due mainly, of course, to the excellent level of illumination in all areas.

‘(*b*) No. . . .

‘10 . . . I would particularly draw your attention to the Philips “light wall” in the restaurant.’

This last reference is to a large mural device producing a slow but continuing change of pattern and colour. This attention to visual interest during breaks from work is shared by many managements and their architects. Another response to question 9 (from Michael Laird & Partners about the Devro Ltd. factory at Moodiesburn, Scotland) says: ‘There have been no adverse reactions from staff, but this may be due to ancillary accommodation such as cafeteria giving an open view to the outside.’ The idea of ‘access to view’—paralleling that of access to quiet—is mentioned increasingly in discussions of working environments.

It is clear that people like windows. But it is also clear that, where there are good reasons for not providing them, and as long as the interior is large enough to give the opportunity of relaxing the eye muscles by focusing on some relatively distant plane, an acceptable visual environment can be designed for an enclosed working space. The element we are likely to see developed is that of variety; this does not mean imitation of natural effects necessarily, but some programmed change based on natural rhythms.

Although the difficulties of accurate costing of lighting or the estimation of its economic value are recognised as considerable in all fields, it might reasonably be thought that, in industry, records would exist of wages, raw material costs, installation expenses, energy bills and so on, to an extent that would encourage economic analysis. It is true that more data are probably available in this sector than elsewhere, but the problems remain daunting.

Chapter 24

Health and welfare buildings

The layman thinks of hospitals primarily in terms of the ward situation, and it is certainly in the patients' rooms that the major difficulties in hospital lighting arise. They are not primarily technical, but much more a matter of human judgment bringing together the needs of patient and medical staff. Whether consciously or not, the patient in hospital is always under observation, and his appearance is, even though to varying degrees, a matter of clinical concern throughout the building. One understands, therefore, the demand from medical staff that the colour rendering of the light sources should be not only good but also standardised; hence the study carried out by the UK Medical Research Council (MRC) about ten years ago to find the most suitable tube colour. The results appear in an MRC memorandum (Reference 2.53). The professional observers' preference was for fluorescent lamps of high colour-rendering quality in the 4000K region—types such as Trucolor 37 and Kolor-rite. However, the patients were not asked for their views—and this comment is not merely cynical, because conflicting criteria must be recognised. Subjectively, the patient needs a sympathetic and reassuring visual environment, one with domestic rather than institutional associations. The colour temperature of the preferred tube types was chosen under test conditions with a higher illumination than is typical of ward levels. This is important, because it appears to ignore the fundamental relationship between quantity of light and the psychological acceptability of near-white source colours. It is difficult to suggest precise levels of illumination at which sources of different colour temperatures become acceptable, because the picture is complicated by the colours used in surface finishes (and indeed by objects in the space) and by expectations associated with the kind of interior. But an average level of 100lux or less is not consistent with a 4000K source producing a happy impression. What is needed at these illuminations is tungsten, or its fluorescent equivalent, and this has the added advantage of domestic associations. It is true that, in a ward with general fluorescent lighting from the medically preferred tubes, the presence of tungsten lamps in the bedhead fittings brings warmth to the total impression, but the fact that these lamps are tolerated —as are purely tungsten-lit wards—makes the objection to, say, Softone 27 difficult to accept. There is no doubt which the patients prefer. In treatment rooms and the like, where patients spend relatively short periods, and where they are examined under higher illuminations, the case for Trucolor 37 or Kolor-rite is a good one, but any insistence that no other type of fluorescent lamp be used in the building reduces the matter to one of convenience in maintenance. If such a ruling must be observed, the answer is not to use fluorescent lamps in the wards at all.

The extensive 'Nightingale'-type ward has largely been superseded by spaces for four or six patients; the essence of the lighting situation is shown in Fig. 4.6.

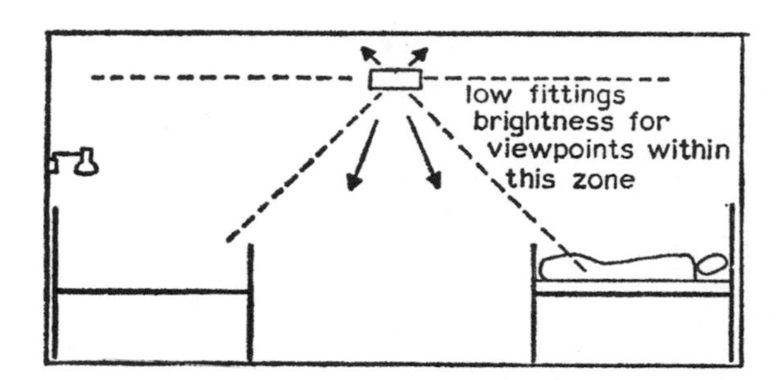

Fig. 4.6 Essentials of wardlighting

The desirable brightness conditions can be expressed in terms of limitations on luminance within stated zones, or they may be implied in recommendations about the horizontal illumination produced by the general lighting system at different points. A table in the IES technical report on hospital lighting (Reference 2.12) quotes an average of 100–200lux in the circulation space (between the bedfoot rails) falling to 30–50lux at the bedhead, where the patient's reading lighting, under his control, gives a minimum on the page of 200lux. Where the ceiling is too low for pendant fittings, a combination of recessed fittings on the centre line and some upward light from other sources is acceptable; the latter should give 30–50lux generally, and the central fittings an additional 150lux or so. Matron may not be very keen on indirect lighting in wards, as it is likely to mean upward facing, near-horizontal surfaces on which the accumulation of dust is not obvious. By the same token, open louvers with vertical blades will be preferred to other methods of light control.

There should be no difficulty in arranging for adequate daylighting in a 4-bed ward with the axis of the room, between the facing pairs of beds, perpendicular to the window wall. As the depth is increased to take three beds each side, however, the ceiling height must be increased to a point that is doubtful economically, especially if it applies to the whole floor of the building. A p.s.a.l.i. approach is feasible here, with a permanent supplementary source for the area remote from the window, but considerable modification of the normal louver or diffuser system is needed to produce comfortable conditions for people lying in bed looking upwards. One possibility is to have the visible brightness largely on the end wall, as a direct balance to the window luminance, even if it is produced by sources within the ceiling space above some asymmetric light-control medium. The permanent supplementary artificial lighting of deep hospital wards is the subject of a paper published in 1970 describing a study at the UK Building Research Station (Reference 2.54).

The race-track-corridor form is appropriate to large hospital buildings, with the patients' accommodation around the perimeter, daylit or using p.s.a.l.i., and the inner zone occupied by treatment rooms, dispensaries, storage, vertical circulation, services and so on—shown diagrammatically in Fig. 4.7.

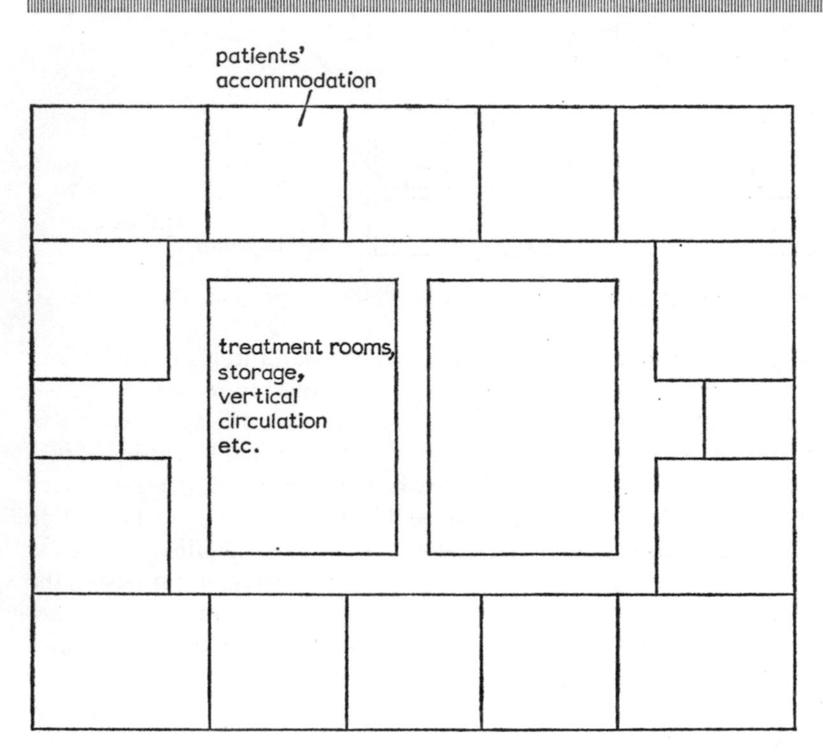

Fig. 4.7 Diagrammatic layout of typical floor in deep multistorey hospital block

The plan has been arranged with the 6-bed rooms at the corners to permit supplementary daylighting; but, whether the perimeter has daytime electric light or not, one enters the corridor from what appears to be a daylit zone (in some layouts, the circulation space may not be completely partitioned from the wards, but this merely increases the difficulty). What emerges from thinking about moving from ward to corridor to central zone at different times of the day is the conclusion that hospitals represent the most positive example of a principle that in essence applies to all buildings, namely that the balance of brightnesses should be considered for the three main states of adaptation: daylight hours, evening, overnight.

During daylight hours, the simultaneous presence of a high visible external brightness, or the recent experience of a daylit interior, means a state of physiological adaptation that makes it necessary for circulation and internal zones to have related brightnesses. Psychological expectations confirm this need (Chapter 14). In the evening, the wardlighting levels are on a domestic scale. Where work is continuing in the core regions, the visual tasks involved are likely to demand higher levels, though not necessarily those mentioned during the day. The corridor spaces must be related to both. After 'lights out', there is need for working light at nurses' stations and for a maintained low level for observation—the IES Report recommends 0·1 lux generally over the ward area (1 lux in children's wards). The circulation lighting must permit safe and confident movement but also allow ready adaptation to these low levels as staff enter the rooms where patients are sleeping. A paper in *IES Transactions* for 1966, 'The lighting of compact plan hospitals' (Reference 2.55) examines these problems in detail, and makes numerical recommendations—these were later refined in the technical report (Reference 2.12).

The principle justifies repetition: in all building types, we should start with the assumption that electric-lighting needs will be different during daylight hours, in the evening and overnight. The extent to which a simpler approach is appropriate will then emerge—but this is the way round that it should happen.

The heightened sensitivity of the sick makes it particularly important to avoid annoyance from choke hum or from flicker in hospitals and similar buildings. Good-quality control gear is essential, and attention to how it is mounted. Flicker from fluorescent lamps is of two kinds, a 100 Hz variation in light output over the whole tube, and a 50 Hz fluctuation near the electrodes at each end. Human sensitivity to flicker varies considerably from person to person (so it is unwise to dismiss complaints dogmatically), but few people can detect directly the 100 Hz change. It may, however, cause stroboscopic effects, as when some moving object appears in a series of discrete positions; special twin-lamp circuits can change the phase of the variation in one tube relative to its partner and so smooth the combined output, but such techniques seem to be used less today than they once were, probably following the increased 'afterglow' from the phosphors in modern fluorescent lamps. Warm tube types generally show less flicker than the cooler alternatives—another reason perhaps for using them for wardlighting. The main precaution against flicker at 50 Hz is the fitting of plastics sleeves, about 40 mm long, to cover the ends of the tube. The complete answer lies in high-frequency operation of the fluorescent lighting (Chapter 13), and it may be that we shall see this introduced in hospitals before other building types.

Many specialist areas in medical buildings have demanding visual requirements; operating theatres are an obvious example. Close liaison with the client is the only way of establishing performance needs and preferences, and it is vital that the actual users be represented as well as management and engineering staff.

Clinics, nursing homes and sanatoria present similar problems, though they may be less critical than in a general hospital. In infant and child welfare buildings, and in homes for the aged, a domestic feeling is essential, and may justify exclusively incandescent schemes (on the other hand, in special baby-care areas, the use of the MRC-approved fluorescent lamps may help in revealing early jaundice). Mental hospitals clearly present specific problems of safety and security with all electrical equipment. There is some evidence that the anxious personality is more sensitive to flicker, and it seems possible that direct light on the ceiling at night may make nightmares more likely. A great deal remains to be learnt of the effect lighting can have no mental states, but there is clearly great potential in the idea of positively therapeutic environments.

Chapter 25

Refreshment, entertainment and recreation buildings

It is no easier to answer the general question 'how do you light a restaurant?' than it is to reply to the inquiry 'how do you cook a meal?' The menu must be chosen for the occasion, and the recipes will reflect the resources available, the impression being aimed at, and the personality of the chef. Some underlying principles may, however, be suggested.

The first is that the more light there is, the less time people will tend to take. Suppose one lists a range of eating houses with the typical duration of the meal and a likely illumination level; the result will be something like Table 8.

I have been trying to establish for some years that the figure in the last column should be known as 'Boud's constant', but the world is slow to recognise its value. The principle, however, is important, and not least because the customer's sense of security depends on feeling confident that he understands the nature and status of the establishment—he will not feel relaxed and expansive if he doubts his ability to pay the bill.

The change between lunch and dinner implied in Table 8 need not be limited to illumination. It is yet another example of the need in many buildings to

Table 8 Relationship between time spent over meal and illumination provided

Type of restaurant	Meal duration	Illumination	Product (duration× illumination)
	min	lux	
Snack bar	10	600	6000
Cafeteria	15	400	6000
Quick service cafe	20	300	6000
Popular restaurant (lunch)	30	200	6000
(dinner)	60	100	6000
Sophisticated restaurant (lunch)	80	75	6000
(dinner)	120	50	6000
Night club	240	25	6000

differentiate between daytime and evening lighting. As the level drops, the colour should be warmer, the diversity greater, and the relative brightness of vertical surfaces less. The practical expression of these ideas may involve an installation using both incandescent and fluorescent sources in the middle of the day and incandescent alone (and possibly dimmed) at night. Building the tubes into concealed positions—at the edge of a dropped ceiling or behind banquette seating—avoids any impression of lamp failure when they are not used.

While the importance in a restaurant of a memorable ambience is clear, the main reason people have for returning is that they enjoyed what they had to eat. The primary function of the lighting therefore is to add to this enjoyment directly by making the food attractive and appetising, rather than showing off the decoration or facilitating seduction. The simple fact is that all food and wine, silver and glass look their best in the direct beam from a concentrated source. Cut into a steak under indirect lighting, and it will look as though it has been kept hot for an hour; put a spotlight on to it, and the sparkling reflections in the drops of surface moisture confirm its freshness. Taste is closely related to texture, and it is directional light that reveals texture. The phrase 'soft lights and sweet music' has a lot to answer for. The lights may be low, but some element of sparkle is essential; candles have more than the atmosphere they create to recommend them.

Lighting for food and drink is thus a kind of display lighting, but one where the audience is onstage and the customers are in the shopwindow. The most effective treatment of the dinner must be reconciled with the visual comfort of the diner. In other words, the food and most of the other things on the table must be directly exposed to the brightness of the source, while the people at the table are not offended by high brightnesses within their normal field of view. Two approaches are effective: putting the source as low as possible, or as high as possible. 'As low as possible' means just above the line of sight eye-to-eye across the table. Pendants here are normally possible only if the table layout is known and fixed, though suspending them from track gives some flexibility. An advantage of this position is the considerable angular separation of the directions from fitting to table and fitting to eye. This makes a large change in the brightness in these two directions feasible. One practical solution is a deep metal housing for a reflector lamp; the inside of the rim should have a dark matt finish. The alternative approach, 'as high as possible', tends to mean downlights recessed into the ceiling, again related to table positions if possible. There is now a very small angular separation between the lines to the table and to the eyes of the man at the table, but, in the ordinary way, he is unlikely to look upward steeply enough to be bothered. The fittings in other parts of the room present a low apparent brightness.

Dimming is obviously valuable in tuning the appearance of a restaurant interior, and in allowing adjustment for different external conditions and various occasions. It is usually possible to confine it to tungsten circuits, where the colour change is appropriate. Patches of warm light seem right too in a bar;

70

Plates 70 & 71
Maintenance from above avoids disruption of work industrially. This plant has one of largest single-span printing halls in Europe

71

Plate 72 High-bay mercury fittings mounted on roof structure above level of travelling crane in Japanese rolling mill

Plate 73 Units for high-wattage mercury lamps recessed into roof structure of windowless production area

Plate 74 Continuous rows of recessed fluorescent fittings in Dutch spinning mill

Plate 75 Standard practice for light industry: trunking on underside of roof structure carries slotted-trough reflector fittings for twin 65 W tubes.

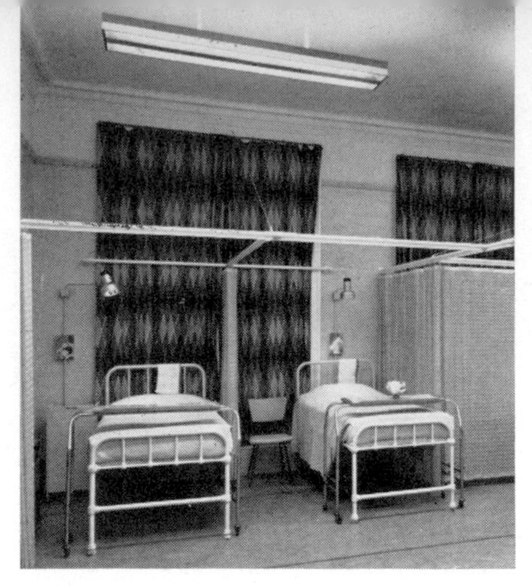

Plate 76 Fittings for ward lighting. General illumination from louvered fluorescent pendants designed to give generous upward light and low brightness from patient's viewpoint, and to present few surfaces on which dust may collect. Bedhead fitting for g.l.s. tungsten lamp has controlled degree of adjustability

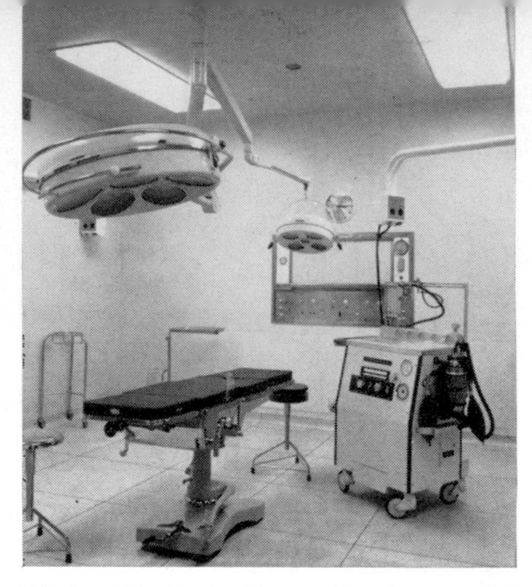

Plate 77 Operating theatre with general light from sealed recessed fluorescent fittings with opal diffusers. Working light is from specialist equipment

Plate 78 West Herts. and Watford Postgraduate Medical Centre, before occupation. General light from batten fittings for fluorescent lamps set in custom-made troughs. Emphasis lighting uses flexible system of spots in surface-mounted tract. *See also Plates 67 and 68.*

any drink with bubbles benefits from sparkle in the lighting. Level, diversity and colour are often among the differences between public and saloon bars—a subject worthy of sociological study. Special lighting is sometimes provided for a dartboard, but other pub games should be remembered.

Kitchens in refreshment, entertainment and recreation buildings range from the domestic in scale to the food factory preparing banquets for hundreds of guests. Waiters moving from the public rooms to serveries should not be subject to visual shocks, and the views the guests get as doors swing open must be remembered. In any case, it is desirable for food to be prepared under lighting conditions related to those where it is served. The delicacy of some of the visual tasks and the need to combine speed and safety may imply an illumination of anything between two and ten times that in the dining room, but consistency of tube colour is worth while (or near consistency: Softone27 might give way to De Luxe Warm White, but not Warm White). The layout of lighting fittings will depend on that of the catering and ventilation equipment. If extract hoods are prominent, a modest overall fluorescent scheme may be supplemented by tungsten fittings within them and so related to the main areas of attention. All lighting fittings used in kitchens should have sealed lamp enclosures to exclude steam and insects, and finishes should be suitable for areas of heavy condensation; exterior types may be appropriate, such as plastics bulkheads within the hoods.

The theatre is a significant building type for the lighting designer in a number of ways. Historically, it was probably the first to exclude daylight consciously, so that the reaction of the people within it could be more completely controlled (in shops, this has happened relatively recently). Then a theatre houses a very positive social assembly; even today, a sense of occasion remains. Whereas many of the interior spaces have apparent equivalents elsewhere—bars, restaurants, offices, workshops—being in a theatre gives them a special quality. All drama has visual elements (though the extent to which it can be regarded as a visual medium varies considerably with the style), and it is on the stage that we find the most extended tradition of lighting for effect. Though the tactical use of stage-lighting equipment is the responsibility of the production staff, the building designer, consciously or not, sets the limits within which they can work. In its turn, their work should be of continuing interest to him, since it explores many of the unquantifiables of lighting effect.

In the traditional Shaftesbury Avenue and the provincial touring theatre—still the basis of most local-authority thinking—the division between front-of-house and backstage is expressed in the building by a fire wall; the proscenium opening is a hole in this wall, which can be filled by the safety curtain. Originally, all the stagelighting equipment was upstage of the proscenium line (with the possible exception of one or two cumbersome carbon-arc 'follow spots'). General illumination of the acting area was provided by battens, i.e. linear assemblies of small floodlights, mostly suspended overhead. An actor moving downstage was lit from above, and from the sides; the closest to frontal light came from the footlights—battens along the bottom of the picture frame.

Too close an approach to them had to be avoided if a patchy effect was not to follow (particularly if the lamp compartments were arranged, as they commonly were, in three circuits corresponding to different colour media in repeating sequence). So footlights tended to be regarded as reinforcing the barrier between actor and audience created by arch and wall. Their disappearance followed the introduction of light sources powerful and flexible enough to provide frontal light from the auditorium; this happened to come at roughly the same time as the movement towards 'breaking out of the proscenium arch'.

Some form of endstaging remains the norm in the theatre, though it may involve an acting area at one end of a single interior housing audience and player, rather than a view through a hole in a wall. Whatever the variations in detail, this is essentially 'one view' theatre; the production is designed and lit for people sharing a view from one side. In placing the actors, this implies a 'face front' convention softened to a varying but limited degree by the demands of realism. To the extent that generalisation is possible, the lighting of the actor's face is most effective if the main beams reaching it come from about 45° above the horizontal, and, in plan, at a similar angle to the audience's direction of view—which we may take as the longitudinal axis of the auditorium and stage. It is clear that the actor has to retreat some way upstage for sources over the acting area itself to be effective, and that, when he does, those just behind the opening will matter most. Thus it is that, in recent layouts for endstaging, as many as half the total number of circuits are front-of-house (i.e. for lanterns in the auditorium area) with perhaps half the balance for the no. 1 bar or barrel over the front of the stage. The remaining lighting, on further barrels, portable stands etc., is mainly for set and cyclorama. For the onstage area, the long tradition of flying (i.e. raising and lowering scenery and equipment) makes access to the lanterns relatively easy, though bridges across the acting area as well as the galleries running fore and aft each side can be valuable; but, in the auditorium, access to the stagelighting fittings merits exhaustive consideration. One is tempted to declare dogmatically that there must be permanent and convenient access to every lantern in its operating position for fixing, lamping, aiming, focusing, fitting colour media and so on. On the smallest scale, working from a ladder may be acceptable, provided that thought has been given to the security of the ladder at both ends; but, beyond this, some system of bridges is essential. The extent to which they are exposed or concealed reflects the style and size of the auditorium. In a small modern theatre with something of the drama workshop feeling, bridges forming part of the exposed steelwork of the roof structure can be completely acceptable. On the other hand, suspended ceiling panels can screen the bridges and associated equipment from most viewpoints. They may have an acoustic function as well as visual justification; at the Abbey in Dublin, Eire, they are on raising and lowering gear, and carry the auditorium lighting, which can thus be serviced from seat level (note the jargon: stage lanterns mounted in the auditorium are 'front-of-house lights' or f.o.h. circuits, whereas auditorium illumination is from the 'house lights').

As an example of the interrelation of beam direction and mounting position, consider the stage and 500–600 seat auditorium shown in diagrammatic plan in Fig. 4.8.

The actor at A on the prompt (audience's right) side of the stage would tend to be lit from the left-hand side of the auditorium. Bridge 1 provides a range of suitable positions as far as the angles in plan go. Fig. 4.9 shows a section along the centre line of the auditorium, but this is only indirectly relevant.

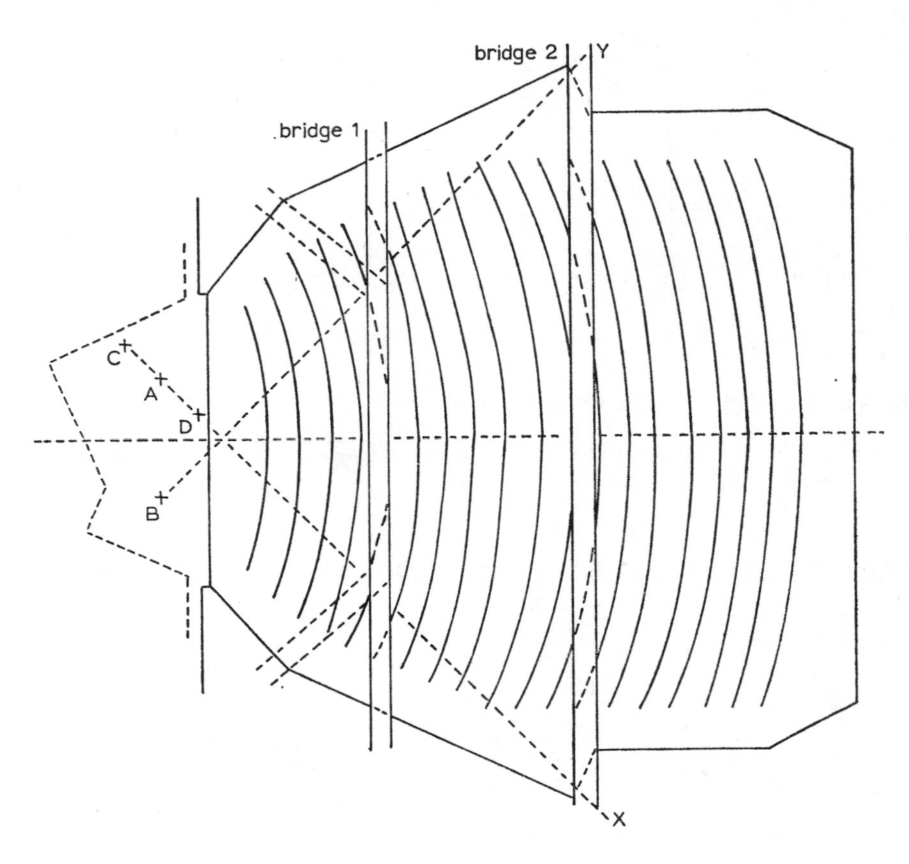

Fig. 4.8 Lighting bridges crossing auditorium

It is the vertical plane through the line AX (Fig. 4.8) that is important, and this is the section that must be considered.

It appears from Figs. 4.9 and 4.10 that the normal longitudinal section that is likely to appear in architectural drawings makes the beams from the two bridges seem steeper than they are in reality. They thus tend to suggest that the source F is more useful than it is; Fig. 4.10 shows that its beam is at less than 30°

to the horizontal, which will mean a rather flat effect and the possibility of troublesome shadows being cast. Equally, the longitudinal section makes S look rather close to the stage; Figure 4.10 shows that, even when the actor advances to D, the beam from S to his head is still just below 45°. The drawings reveal that bridge 2 has a limited usefulness, except for special purposes like follow spots

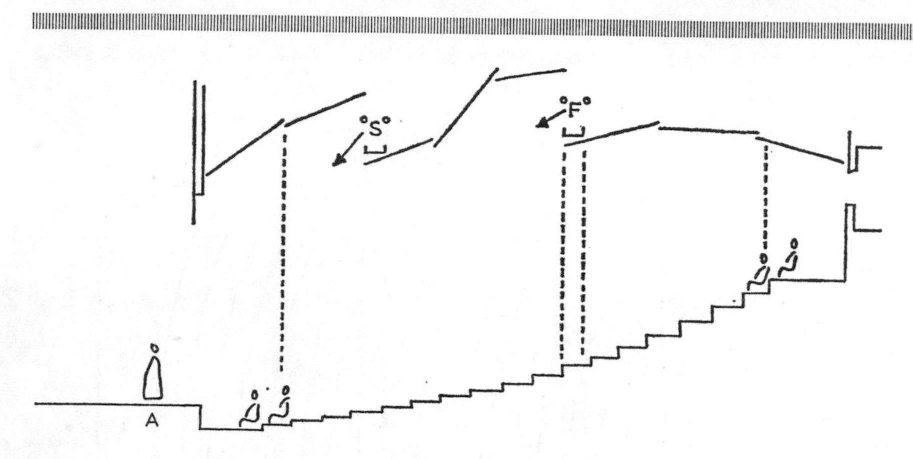

Fig. 4.9 Longitudinal section of auditorium

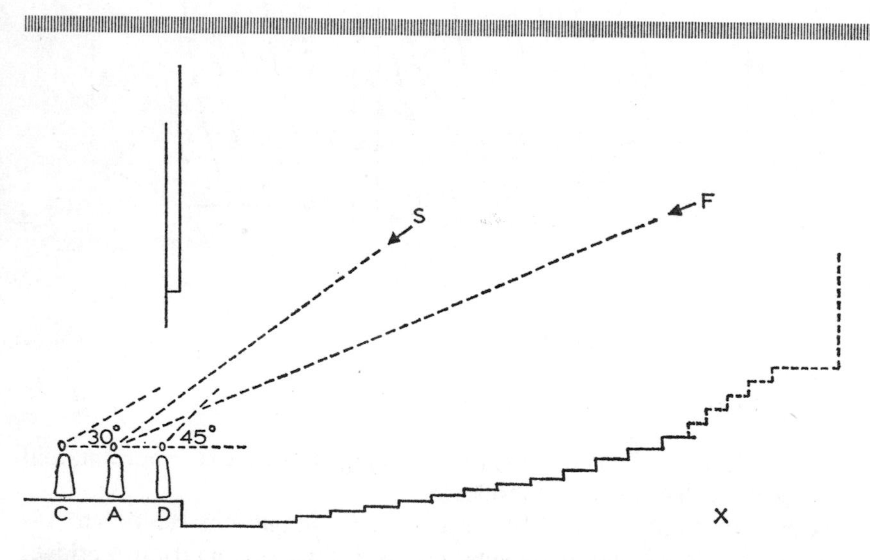

Fig. 4.10 Section of auditorium in vertical plane through AX

or projection. A series of vertical mountings over the entrances close to X and Y would meet most needs for sources further away than bridge 1—although, in itself, access to the overhead space in this part of the auditorium could, of course, prove useful. Bridge 1 could be improved, if it were structurally possible, by making each end an angled return (broken lines in Fig. 4.8).

A small forestage makes little essential difference in a theatre rooted in the one-view tradition, but, as it increases in size toward a thrust stage, we reach a dangerous transition zone between one-view and multiview. It is better to regard 3-sided or thrust staging as a variation on the complete arena ('theatre in the round') than as a development from the one-view situation. When the acting area is completely surrounded by an audience, there are severe constraints on beam direction if glare and distraction are to be avoided. Since there is no 'front', the actors have no reason not to face each other directly. The beam at about 45°, which lights the face of an actor at the edge of the stage but looking inwards, inevitably overshoots and can cause problems in the first row of audience behind him—either visual discomfort or making it too conspicuous. The difficulty can be lessened by increasing the horizontal separation between the acting area and the first row, by using floor level for the stage rather than raising it (or, better still, a lower level if the rake on the seating is adequate for good sight lines), and by inclining the beams closer to the vertical. Some 60° to the horizontal should be the maximum, however, for, apart from the modelling of the face, it is important for vitality—and, oddly, audibility—that direct light reaches the eyes. Most arena theatres are small, and will need only three or four lanterns over the acting area. The rest of the lighting is over the heads of the audience and directed towards actors on the further side of the stage; profile spots will be used for their positive cutoff. Some layouts tend to distribute the sources fairly uniformly around the arena, as though for a circus or show jumping, but there is a good case for concentrating them in, say, three zones separated by some 120° in plan. Unless there is a realistic inconsistency with the play's content, one group could be without colour media and the other two lightly warm and cool, respectively, to bring some shape and coherence to the effect. It is true that open staging, by its nature, cannot offer the pictorial possibilities of the picture frame, but it is less than just to claim that 'mere illumination' is all that is possible.

Thrust staging adopts related techniques, though probably on a larger scale, and is in a sense easier to design lighting for, since it has a fourth side without any audience. But if 3-sided theatre is to follow Polonius's dictum 'to thine own self be true', the architect must remember, with producer, players and lighting designer, that, though the view from the sides is different from that along the centre line, there is no justification for its being worse in any way. Theatre design and production planning should always proceed on the assumption that all the seats are to have the same price.

If these brief comments—on some aspects of theatre design, the provision to be made for stagelighting equipment, and the way equipment is used—have shown how closely interrelated these matters are, they have achieved their purpose. More than that is impossible here; there has been no reference to the equipment itself, lanterns or control systems, or to many other matters that affect stagelighting planning. Specialist advice is obviously essential, and I should mention the UK Society of Theatre Consultants, the Association of British Theatre Technicians, and the Society of British Theatre Lighting Designers. Among the more important books are Frederick Bentham's 'The art of stage lighting' (Reference 2.56) and Richard Pilbrow's 'Stage Lighting' (Reference 2.57). Rank Strand Electric Ltd., the major UK supplier of stagelighting equipment, publishes a quarterly magazine, *Tabs*, and the useful booklets 'Stage planning' and 'New theatres in Britain'. Playhouses in Sheffield and Colchester have been described in recent issues of *Light and Lighting* (References 2.58 and 2.59).

Auditorium lighting must reflect the general style of the building—drama studio or opera house? The temptation within the modern idiom is to respect the importance of the stage itself so much that the auditorium is left aggressively austere; it should, at the least, stimulate expectation. The sparkle of small exposed sources and traditional crystal glass was not without its point. Where the interior is used for more than one type of activity, a range of effects becomes desirable in the auditorium as well as on the stage. Particularly if it houses conferences that extend over several days, it is important to be able to produce a higher and more uniform illumination than the punctuated scintillation appropriate to evening entertainment. In a concert hall, the maintained level during performance must allow miniature scores to be read, but the lighting that stays on should draw no attention to itself. Another difference in the musical world is that the performers accept glare less readily and must be able to read their parts easily. If 45° is the ideal elevation of beams on to actors, it is perhaps 60° for musicians—and a little more for politicians.

Complete auditorium lighting, in addition to conventional house lights and separate maintained lighting during film or play, should include lighting for cleaners, luminous exit signs, and emergency provision. The latter will be influenced by the requirements of the local authority, which vary from area to area and are most demanding, reasonably, for places of public assembly. At the 1972 IES national lighting conference at Warwick, England, the confused state of recommendations on emergency lighting was lamented; efforts directed at the achievement of some degree of national and international standardisation are to be encouraged. In the UK, BS Code of Practice 1007 refers only to cinemas; exit signs are covered in BS2560 and BS4218. The most recent useful contribution to the literature on the subject has come from a trade association (Reference 2.60).

The basic strategic decision with sports halls, swimming pools and gymnasiums is whether to have daylight or not. Here, as in other building types, overglazing in the recent past has tended to produce an exaggerated reaction. Glare from direct sunlight, or even where this is screened, from the brightness of the sky seen directly or reflected in floor or water, has led to clients demanding blind boxes. Airconditioning engineers have welcomed this isolation as a help in keeping internal temperatures down in summer. However, in a project being planned at the time of writing, a system of indirect daylight has been proposed. This means that there will be no direct view of the sky from any position that a user of the building normally reaches; but natural light entering via linear barrel rooflights is reflected back from catwalks and baffles beneath (they also carry the electric lighting); it not only brightens the underside of the roof to reduce the contrast between lighting fittings and their background, but provides an element in the internal field of view that responds to outside conditions. If nothing more, this proves that the interior is not underground. Many lighting engineers can see little point in indirect daylight of this kind, but I believe we shall see more of it.

What is probably the major special visual problem in sports halls is revealed when players follow the flight of a ball overhead. It moves across a background of ceiling and upper walls, and so one of contrasting brightnesses presented by lighting fittings (or rooflights) and the areas between them. A uniform luminous ceiling would be an expensive solution, but, in addition to its problems of installation, maintenance, and mechanical protection, it would not be ideal visually, as the extremely diffuse effect would result in weak modelling in the space beneath. It is probably better to aim at spreading the brightnesses presented by the main light sources over, say, 10–20% of the visual field overhead, and making sure that the intervening areas have a luminance that does not present too great a contrast—as an arbitrary rule of tumb, one might suggest 'not less than one-tenth'. Any brightnesses greater than that of the main source—as from tungsten fittings added for stronger directional effect—should be within the main-source area. In some such way, a controlled gradation of luminance can be established. Sophisticated calculations are unnecessary, since no very precise criteria exist, but some mockup that can be assessed subjectively will be valuable, if only to show the client what he can expect, since numerical statements in luminance terms mean little to the layman (and not much to the specialist perhaps).

Sports halls and similar interiors have attracted the attention of advocates of increased use of discharge sources indoors. If you regard such an interior as basically industrial, there is a case that can be made in terms of the small number of powerful sources representing a simpler scheme, easier and cheaper to install and maintain. However, although to some extent this is a personal reaction, I remain unrepentantly in favour of tubular fluorescent lighting as the main source in this context. It offers better colour, less flicker, and immediate restriking after any interruption in the electricity supply; but more than this, the physical size of the tubes is a great advantage visually, in achieving the spread of source brightness that is so desirable. This aspect has already been discussed as it applies in industrial buildings (Chapter 23), but the point bears repetition: a smaller luminous area implies a lower BZ number to meet a given glare-index criterion, and this more narrowly concentrated downward lighting means a less satisfactory visual effect. This is not to say that high-pressure lamps might not successfully be used in sports halls to supplement fluorescent tubes; mercury types could provide an upward flood for ceiling brightness, or just possibly high-pressure sodium lamps could be used instead of incandescent lamps for the directional addition from above.

In all buildings of this kind, access to the equipment needs careful thought at the design stage, and this is particularly true of swimming pools. The solution that can almost be assumed here is that servicing will be carried out from above. The ceiling should have sealed portholes or panels through which light is projected from equipment outside the pool atmosphere with its high humidity and corrosion potential from the chlorine treatment of the water. Where spectators are normally present, underwater lighting can be effective; this again usually depends on standard lighting equipment mounted in a service corridor with portholes rather than on submersible fittings. For exhibition diving, spotlights can be trained on the points in the air where the divers seem to hover momentarily.

The IES Technical Report on sports-lighting (Reference 2.7) is the most important reference here, and new installations are reported regularly in lighting journals.

Plate 79 Light sources on tables in restaurant reduce scale and concentrate attention. Here lamps are beneath table-surface level

Plate 80 Low-brightness recessed downlights create emphasis on tables. Light-coloured cloths and fairly reflective carpets, together with perimeter fluorescent, prevent too strongly vertical effect on faces

Plates 81 & 82 Contrast in sports-hall lighting. Main hall of Wiener Stadthalle, municipal sports centre in Vienna (Plate 81) presents more dramatic picture for spectators, while multipurpose sports hall at Eindhoven gives better visual conditions for players

Plate 83 Discharge-lamp lighting for Sunderland Football Club training gymnasium

Plate 84 Squash-court lighting: metal angle reflectors for playing wall and diffusing plastic trough reflectors for remaining fittings

Plate 85 At Royal Commonwealth Pool, Edinburgh, access through roof void permits servicing of fittings from above

Plate 86 Ceiling of meeting room at Kyoto International Conference Hall, Japan

Chapter 26

Sacred buildings

The older approach to churchlighting put great emphasis on the fittings themselves. The forge worker in Ronald Blythe's 'Akenfield' remembers that 'the cathedral lights took us a year to make and we had to work until midnight to finish them off'. The medieval tradition of craftsmanship must be respected—no cost/benefit analysis here. Without this dedication, sacred buildings used to be lit by items selected from a small section of the manufacturer's catalogue headed 'churchlighting fittings'—mechanically reproduced Gothic ornament and little attention to light distribution. More than most could these rows of pendants be accused of polluting the interior they feebly illuminated.

Today the approach is usually different; the fittings are inconspicuous and the emphasis is on their effect—a sort of interior floodlighting of the building, with the 'working light', for reading the prayer book or whatever, appearing almost incidentally. A key installation in the progress was the relighting of Gloucester Cathedral in 1958, designed by J. M. Waldram and described in a paper he gave to the IES the following year (Reference 2.61). It was based on the 'designed appearance' method that Waldram had developed and which involves fairly sophisticated calculations. But as F. P. Bentham observed in the discussion after the paper, '. . . the method and the calculation are as nothing compared to the original decisions on level and direction. . . . We must have the inspiration of the artist here, an inspiration which has its genesis in a loving and devout study of the cathedral and its use. . . . This dramatic lighting has its origins in the theatre and, where effective there, it is the result of the same saturation . . . and study. . . . The dimming of some of the circuits may be an aid.'

Much of this comment could happily be applied to a wider range of interiors. Outside the purely functional requirements of marshalling yards or standardised factory interiors, and system of lighting design that stresses calculation has questionable priorities. Numbers are useful in establishing, say, the range of loadings within which one is working, or checking that a luminance relationship is more or less as one anticipates; the sums are essentially marginal, an aid to design but not its central thread. The designer must be confident enough in his handling of the calculation techniques that have been developed to treat them as servants, not masters. Except in the simplest, routine situation, lighting design should begin with a period of apparently unproductive reflection, 'saturation and study'. And as Joh. Jansen has pointed out, separating the components of a scheme into circuits with individual dimming not only permits a variety of effect but also removes much of the need for precision in prediction techniques.

In a modest scheme for a local parish church, the lighting was broken down into four groups of circuits:

(*a*) chancel lighting including directional effects and emphasis at the altar
(*b*) restrained upward lighting to the dark timber roof of the body of the church
(*c*) main downward illumination for this area
(*d*) aisle lighting.

Dimming was provided only for *c* and *d*. The upward flow from *b* depended upon 60 W g.l.s. pearl lamps with clip-on reflectors in simple holders, and was just enough to avoid the sense of characterless darkness overhead. Adjustment of the levels for *c* and *d* makes possible a wide range of effects in terms of the balance between the chancel and the rest of the church, and the opening-up or enclosing effect of taking the relative brightness of the aisles up or down. Even when dimming is not included, it is important to arrange the switching for lighting effect and not simply for installation convenience.

Bentham's mention of the theatre in relation to churchlighting has every historical justification, but it points the need for sensitivity over the degree of visual drama. As a long-relapsed Anglican, I feel some understanding of what is appropriate in the Church of England, though even here it is important to be clear whether the local tradition is 'high' or 'low'. But in relighting a Christian Science Church some years ago, or designing a scheme for a mosque a little later, I felt, and I am sure rightly, that lumens and watts were less important than some idea of the feeling of the ritual—or even whether the term could be applied. Inquiry must proceed with delicacy if a detached, 'comparative religion' approach is not to seem offensive to the committed.

The size of sacred buildings and the esteem in which their fabric is held throw into relief problems of access and fixings: and cable routes can be as much a problem as mounting the fittings. One can offer little generalised advice but a modest booklet from the Council for the Care of Churches is helpful (Reference 2.62). An unhurried consideration of all the implications of the particular case is called for, but the problem must be recognised from the outset as one of the essential constraints. It is a rather tedious and discouraging principle, but no lighting scheme can be better than its maintenance.

Chapter 27

Buildings for education, science and art

Electric lighting for afterdark use in a conventional primary school presents few problems of principle, and there is a body of acceptable precedent as guidance. In the UK, the official recommendations are collected on a Department of Education & Science (DES) bulletin (Reference 2.63). but postwar primary schools are the most frequently quoted example of unintegrated design, in that many of them demonstrate the effect of pursuing an aim that is desirable in itself —generous daylighting—without regard for its repercussions on other environmental factors. It was probably a feeling that nothing was too good for the children that prompted the establishment of the 2% daylight factor standard for teaching spaces. But this led to a generation of school buildings in which overglazing resulted in sky glare, a sense of cold radiation from the windows in winter, and, particularly, excessive solar gain on a summer afternoon. Current refurbishing tends to be limited to much older buildings where the flywheel effect of a heavy structure mitigates thermal effects, but, when the time comes to treat some of the buildings of the 1950s, modifications to the windows to reduce transmission of both heat and light are indicated. This will improve the visual balance between inside and outside, but raising the internal illumination can help in this, and it may be a step that can be taken immediately, perhaps simply by substituting fluorescent fittings for the original tungsten.

The 2% daylight factor remains a formal requirement, but a change in the interpretation of the term makes it less demanding; 'daylight factor' used to mean the direct effect of a uniform sky—now it means the total effect of a 'standard CIE overcast sky' and the light reflected from external and internal surfaces. In a recent building that has justifiably received considerable attention, Eastergate Primary School near Chichester, England, some rooflighting was added to restricted (20%) side-window area to achieve 'the statutory 2% daylight factor within the teaching area. However, the planning of the interior was based on permanent electric lighting to allow full freedom of interior use' (Reference 2.64). In fact, the technical observance of the 2% standard may have been a trifle cynical—obeying the letter of the law, as it were—since, as conventionally measured, daylighting from overhead is not numerically compatible with daylighting from side windows. David Moizer identified these fallacies some years ago in reviewing the DES bulletin in *Light and Lighting* (Reference 2.65). Modern school-planning ideas 'call for complex spaces (within which ephemeral working locations are created using mobile furniture) too large to be lit by side windows alone, thus indicating general use of permanent supplementary artificial lighting. Yet, in this bulletin, p.s.a.l.i., although given considerable space, is only permitted for special situations. For single storey designs, where natural lighting is possible, no guidance on integrating rooflighting and side lighting is given except the 2% d.f. It has been recognised for years that one needs about 5% d.f. from rooflights to get room lighting equivalent to 2% from windows.'

A fully daylit school is certainly possible, but its design must give great attention to thermal comfort and brightness control. It is likely, moreover, to depend on an extended perimeter, and the costs this implies should be recognised as the price that is being paid for the daylight and the visual interest of the sequence of spaces. In most cases, however, it is likely that a designed combination of daylighting and electric lighting will be a better deal in total; this seems to be the direction of current development. Traditionally, the assumption has been that school premises are little used after dark, but, with the growth of the idea of 'open', community schools, the high utilisation for adult evening classes and leisure activities makes a reappraisal necessary. The first reaction to 'how should this building be lighted?' is always another question: 'how is it used?'

Since a comprehensive discussion is not possible here, two further references should be mentioned. Mark Wood-Robinson's article 'School lighting' in *Light and Lighting* back in 1966 remains a useful general guide (Reference 2.66), although numerical recommendations have since changed. Then there is the Pilkington Research Unit's report 'The primary school; an environment for education' (Reference 2.67); this unit carried out some valuable work at Liverpool University under Peter Manning's direction, including investigations of single-storey factories (Reference 2.34) and office buildings (Reference 2.68). One last point on classrooms: supplementary illumination for the chalkboard is almost always welcome, provided the reflection of the source from the board is well away from normal directions of viewing.

Educational buildings embrace a vast range of interiors; within the university context, let us limit brief comments to laboratories, lecture theatres, and libraries. Cleanliness and accuracy are conventionally associated with the white-coated, bespectacled scientific worker, and laboratory lighting reasonably reflects this image: a fairly high illumination from sources with good colour rendering in sealed fittings arranged to give a low glare index. However, the emphasis changes with the type of work, both in visual terms—should precautions be taken, for example, about annoying reflections in instrument dials?—and over protection against corrosion or, perhaps, an explosion hazard.

A lecture theatre is an auditorium, but the kind of stimulation it presents needs to be to intellectual alertness rather than entertainment expectation. There is an unresolvable controversy over the desirability of daylighting. A closed-box approach permits economic planning of the building as a whole, and avoids distraction from outside, both aural and visual; there are no blackout problems. It is clear that a well designed electric scheme, particularly if it has considered flexibility, is preferable to the routine provision of windows with no thought given to brightness balance. But some visual contact with the outside world is almost certainly welcome in itself; perhaps this is another case where indirect daylighting is appropriate, i.e. the use of natural light to brighten walls or other surfaces without the sky being visible. In exploring possible types of glazed aperture, it is worth remembering that vertical glass is, in principle, easier to install, maintain and blackout than glass in any other plane. Visual aids are exploited fully in present-day teaching in higher education, so dimming is essential to finding the right balance between screen and notebook brightness for different types of projection—it will also help in producing a different atmosphere when the lecture theatre is used in the evening or for less academic purposes. IES Technical Report 5 deals with lecture theatres and their lighting (Reference 2.5).

There is also a technical report on libraries (Reference 2.8), covering the general case as well as the educational needs. The contrasts are mainly those of atmosphere, but these accord happily with the differences in visual needs in visiting a community building to select leisure reading and going to an academic establishment for a day's work. In the open-access system, bookcase lighting is clearly important. It presents a geometrical problem that recurs in many different parts of the lighting field: how to get some approach to uniform illumination on a vertical surface from sources close to the top or the bottom—cf. other forms of shelf display or, the other way up, close offset floodlighting. Solutions depend to some extent on selecting, or designing, appropriate optical systems for the fittings (linear lamps are better than point sources), but more on adjusting the geometry to make it a little less unfavourable. This may mean long brackets to bring reflectors well out from the plane of the book spines, or inclining the whole bookcase or the lowest shelves. A floor with a reasonably high reflectance obviously helps too.

Memories of the Radcliffe or the British Museum seem to provoke exercises in the monumental when a major new library is designed, and this throws into relief the problem of reconciling lighting equipment and building fabric. One answer is to stress the 'applied' character of the limited 'building lighting' and to depend to a large extent on sources associated with furniture and fittings —Trinity in Dublin, Eire, is an example. The provision of individual lights at reading tables in libraries has the further virtue of suggesting some kind of a shell, or cell, of privacy; the study carrel expresses this more explicitly. However, the difficulty of finding a good position for the local light is not completely resolvable. A row of linear reflectors about 300 mm above the centre line of a table, separating the territories of the opposing readers, can easily produce troublesome reflections, particularly in glossy paper. Where an individual working space is enclosed on three sides, a small housing for, say, a 15 W tube can be mounted on the reader's left, but to the irritation of the left-handed. There is no magic solution; as is so often the case, we must be aware of the difficulty, and do the best we can about it in the specific context.

Fine-art studios traditionally depend on natural north light, with work ceasing at dusk; but, for painting after dark, the coolest group of good colour-rendering tubes (Northlight, Colour Matching, Artificial Daylight—all around 6500 K) are appropriate only if the level is high enough, say 500 lux plus. Usually Kolor-rite or Trucolor 37 (about 4000 K) are better, and produce no shocks when the work is examined the next day; they are acceptable at 300 lux and upwards, and are widely used. The special graphic-arts fluorescent lamp GraphicA 47, with a colour temperature of 5000 K, should also be remembered, though it has tended to be employed more for inspection than for general lighting. Diffusing fluorescent fittings are best positioned close to the windows to produce a distribution resembling that of the daylight; they can be mounted above the opening, vertically at the sides, or even in boxes which swing round like large shutters to cover the window area itself. Portable diffusing boxes on easellike stands have been used. The coherent flow of light from a large source of this kind is often closer to the effect an artist is seeking than the hard-edged modelling from tungsten spots.

In photographic, film and television studios, the creative application of light is the prerogative of the building user; he asks simply for adequate electrical provision and suitable mounting arrangements—though his work, like that of his colleagues in the theatre, has lessons for the designer of building lighting. Installation in new television studios are usually described in the journals—see, for instance, Reference 2.69.

The lighting of art galleries and museums is another of the special topics covered by the invaluable series of IES Technical Reports (Reference 2.14); conservation aspects are among those examined. I have written about some of the particular problems elsewhere (Reference 2.70). Of the many reports of individual installations, a paper worthy of particular notice is that on the Gulbenkian Museum in Lisbon, Portugal, by Ribeiro, Allen and de Amorim (Reference 2.71), both for its text and for the subsequent discussion. This threw up yet again the need to consider the perceptual background to museum and gallery display. As we read in Chapter 4, visual perception involves four constancies, owing in part to retinal mechanisms, but mainly acquired through experience. These are the constancies of size, shape, brightness, and colour, and accurate perception depends on maintaining them. Brightness constancy is preserved only when the range of luminances in the field of view is relatively limited. When contrasts are stronger, perception is no longer unambiguous. This very ambiguity can be effective—its appearance may help us to identify what we mean by the apparently vague expression 'lighting for effect'—and an analogy with prose and poetry can be developed (Reference 2.48):

'Ambiguity is often the key to richness in art. From the punning of

Shakespeare's clowns to the wandering tonality of 20th-century music, from Picasso's simultaneous profile and fullface to Dylan Thomas's "sloeblack, slow, black, crowblack, fishingboat-bobbing sea", it is multiplicity of allusion that distinguishes poetry from prose—plurality of reference in the context of the expectation that has been created or encouraged. In the prose of lighting, ambiguity is to be avoided—a clear account of our visual environment is all we ask. But in the poetry of lighting, absolute clarity in the message is rarely the aim. It is not by chance that the opposite quality —obscurity—means absence of light, that obscurity is the charge persistently levelled against the *avant-garde*, and that the more adventurous lighting designers claim to pursue the creative use of darkness.'

In lighting an art gallery, we have thus to find a balance between the conflicting demands of stimulation in display and accuracy in appraisal. The ideal balance will vary with the collection, the visitor, the occasion, the building; there may be a case for using switching and dimming to change it from time to time. But the main danger is in allowing the presentation to intrude on the object presented, to show not 'sculpture by X' but 'sculpture by X as lit by Y'. The extent to which dramatisation is legitimate cannot be described numerically, and circumstances alter cases. As in other fields—such as churchlighting—where this kind of balance has to be found, the client is unlikely to be able to formulate what is needed, and time spent in discussion is essential if the designer is to feel what is appropriate. What he is looking for may well be ways of providing lighting that is subtly and consciously related to specific exhibits while it appears to be building lighting—some arrangement of sources giving not only the required emphasis and modelling, but also the essential visual context.

Chapter 28

Residential buildings

A central paradox in domestic lighting is that a scheme that sets out to meet functional needs often produces a happier general effect than a scheme with primarily decorative motivation. It must be acknowledged that personal taste and preference find particular expression in this sector, and rightly so. Nevertheless, it is possible to find a framework for design in a simple checklist: position, intensity, distribution, colour, style, control.

Light sources, at home as elsewhere, need to be positioned in relation to the visual tasks, and these can be anticipated with some confidence in most domestic rooms, since economic pressures today demand fairly tight planning. In single-use spaces like bathrooms and kitchens, the position of the major pieces of equipment is predetermined, so we know exactly where light is needed for the face at the mirror or the washing-up. Bedrooms in most housing today are planned on definite assumptions about bed and dressing-table position. Only in the largest area, the living-room, where we need versatility to house a wider range of activities, is there some flexibility of furnishing arrangement, and even here a good relationship between television screen and window often tends to force the seating into a particular L formation. The effect of all this is to reverse the traditional approach to homelighting. Once upon a time, we provided general illumination for a room—the dreaded central pendant—and then asked whether any local supplement was needed for particular areas. Now it is more reasonable to meet local requirements and then ask whether anything more is desirable. The main lighting needs in the rooms of a typical home have been spelt out elsewhere (see, for instance, References 2.72 and 2.73). Examples are given here in terms of the checklist mentioned above.

In a living-room, the position of the lighting points must be appropriate for casual reading in the main seating areas. Over intensity, a rule of thumb—subject obviously to many qualifications—is 20–25 W/m^2 for tungsten lamps, and about one-half of this if fluorescent is used. A typical living-room of 20 m^2 thus needs 400–500 W, which could mean four 100 W lamps or two of 100 W and two of 150 W—it is very often effective to think of domestic lighting fittings in pairs. 'Distribution' reminds us that the light must fall fairly squarely on to book, paper or sewing while a low brightness is presented to the eyes. This means that an up-and-down distribution, such as that from a drum shade, is appropriate in a living room. If the shade transmits a small proportion of the light, colour in its outer surface can be related to others in the room, but the inside should be white, not merely for its high reflectance but also to leave the source colour unaffected for the major part of the fitting output. 'Style' includes the appearance design of the fitting in relation to the treatment of the room generally—very much a personal matter—but also how the fitting is mounted. Notice, however, that it is reasonable to settle where the source is to be in space, its intensity, its light distribution, and its colour and stylistic implications, before deciding whether it is pendant, wall-mounted, or portable (on table, shelf or floor). 'Control' usually means switching. Something should be switched from the door of the room (on a 2-way system if circulation is through it), but, generally, the more circuits there are the more flexible is the effect. Dimming represents, of course, an enormous extension of this flexibility.

In a dining space—whether a separate room or an area within one of larger extent—the needs are more specific: light for the table and for some serving surface, such as a sideboard. The table lighting follows the principles discussed in relation to restaurants (Chapter 23). Position: usually above the table, either as low as possible or as high as possible. Intensity: 100 W minimum, but dimming desirable. Distribution: basically a strong downward beam, but with attention to avoiding glare for those seated at the table. Colour: tungsten on full voltage or dimmed. Style: many possibilities, as long as the table top is exposed to direct light; sparkle from a number of small sources can be as effective as a single powerful spotlight; fixing can be pendant (low) or surface or recessed. Control: dimming already mentioned; switching separate from sideboard light—the latter, incidentally, can be switched off when everyone is happily at the table, or can be chosen to give diffused background illumination as well as 'working light'. Within a combined living-room, switching from the sitting-area lighting to the table can transform the appearance of the room.

In a bedroom, the main visual tasks are at the mirror at the dressing table, or its equivalent, and at the bedhead. Whether anything more is required depends on the character of the room and the spread of light from the local sources. In a single-room, the bedhead light can have a general as well as a local function, but in a double-room a confined distribution is more appropriate. This may mean that the mirror lighting should provide general illumination. It cannot if it is in an alcove, and some ceiling fitting may be necessary to reveal whether a shirt in a drawer is white or grey or cream, but it will be rare for the centre of the room to be the best position for it. The checklist may be applied, as an example, to the bedhead lighting in a single-room. Position: a little above the head of someone sitting in bed. Intensity: suppose the room is 3 m by 2 m, or 6 m^2; then 60 W at the mirror and 60 W at the bedhead is probably enough. Distribution: downward emphasis, but some general spread (as from an opal-glass diffuser open at the bottom). Colour: since this fitting provides the major part of the room's general light, it should be close to white. Style: a wall bracket with a backplate in a wood appropriate to the furnishing can have metalwork in a finish that is again consistent with general treatment. Control: 2-way switching from door and bedhead.

In a small modern bathroom, 2m by 2m, the wattage rule gives 40W fluorescent or 80W tungsten. The first might be a single 1·2m tube in a wall-mounted diffuser over the mirror, and the second could be two 40W g.l.s. lamps in a pair of pendants at eye level or just above and at each side of the mirror over the basin. The main spread of white light must be sideways towards the face; a pull-cord switch inside the room is probably better than a conventional wall-plate type outside.

As in most other building types, if a scheme is developed on a room-by-room basis, it should be regarded as tentative until the possibilities of rationalisation have been explored. It is obviously advisable domestically to reduce the variety of replacement lamps as much as possible, and the repeated use of one kind of fitting can constitute a welcome unifying thread running through the house.

Exterior lighting should be more than just an afterthought. It is clearly absurd to have to wait for daylight to be able to take rubbish to the dustbin, but, in a high proportion of existing houses, this must still be the case. Outside lighting has a special importance for a patio or similar area with extensive uncurtained glazing, transforming a blackhole into a floodlit court—probably no more than a single 100W PAR38 floodlight is necessary, in a simple external holder. A higher brightness is needed for areas intended to be seen from within the lighted house than is necessary when one is outside; a fitting with a 15W lamp has quite a significant effect amid the surrounding darkness.

Particularly in large-scale housing, architects may feel that the lighting of the individual unit is a matter for the tenant. It is clear, even in traditional terms, that the necessary electrical provision can be made successfully only if it rests on some understanding of its likely and desirable use; but, beyond this, there is the argument that the tight design in most new domestic spaces makes it possible and advantageous to build in standard basic equipment. Housing seems perpetually to be scarce and expensive, and the comparison with motor-car production is sometimes made. It is certainly misleading to push the analogy too far, but there is a parallel between the 'built-in' headlights and taillights that meet basic needs and lighting in a house that might be supplied with the bricks—similarly integrated. The owner then expresses his individual needs and fantasies with applied equipment, fog lamps and the rest. At home, he would buy spotlights and portable fittings—the trend in over-the-counter sales of homelighting equipment seems to be in this direction. A question-mark hangs over the future of industrialised construction, but housing is the most likely field for successful application of this approach, and the factory-made wall or ceiling panel can easily incorporate fully integrated lighting.

Development of these ideas will make it desirable to have some standard homelighting specification for reference. A proposed 'code' of this kind was formulated by Derek Durrant, the UK Electricity Council's senior lighting engineer, in a recent article (Reference 2.74).

The pressures that have led to commercial buildings becoming deeper (with results that have been explored in Chapters 14 and 15) apply in housing too. The question that arose in the early 1960s was whether there was a domestic equivalent for the no-daylight or supplemented-daylight zone in the deeper office block. The quotation that follows is from an article that appeared in 1964 (Reference 2.75; see also Reference 2.73):

> 'The service core of a flat or maisonette—or a terrace house—is its bathroom and kitchen. The bathroom can be without daylight, and the kitchen, while enjoying a view of the outside through a window across a living area, needs supplementary illumination. In a unit on two floors, bathroom and kitchen would be in the middle of the block, one above the other, together with any storage space. The essential high density requirement of narrow frontage per unit and considerable depth is achieved by needing exterior walls only for living rooms and bedrooms.'

In the National Building Agency book of generic plans for 2- and 3-storey houses (Reference 2.75), the 52 basic plans show 13 with internal bathrooms and eight with kitchens away from external walls. However, what has been inadequately recognised in these developments is that electric lighting needed during daylight hours should be considered separately. It is not enough just to switch on the evening lighting. The IES Technical Report on daytime lighting (Reference 2.4) says: 'If the deeper plan permits the reduction of frontage for a given amount of accommodation, and so a greater number of units on a particular site, it seems not unreasonable to expect sufficient and suitable electric lighting to be provided to help the tenant to achieve good visual conditions by day.' The most important factors that distinguish good daytime electric lighting from an evening scheme are a higher illumination and a cooler light-source colour. It is also relevant to aim at greater uniformity and relatively higher brightnesses on vertical surfaces. All of which suggests a fluorescent daytime supplement of some sort. The room in question can have a 'day–night' switch as well as one for 'on-off'; or the circuits might be arranged with a photocell to limit the availability of the fluorescent component to daylight hours (it would be important to be sure that the tenant understood the system). Perhaps, in this latter case, the daytime supplement could be provided at the landlord's expense.

It was suggested in Chapter 24 that, in all building types, we should start with the assumption that electric lighting needs will be different during daylight hours, in the evening, and overnight. We have seen now that, in some types of housing, the daytime case is important; there is just as good a case for a separate, maintained overnight scheme that avoids the necessity of switching on the evening lighting when one is moving about the house during the small hours. Under the adaptation conditions that then apply, very little light is needed, and two or three fittings for miniature tubes can be distributed at low level in the circulation space and bathroom; the consumption is so low that there is little need to switch them off during the day, and continuous operation gives very long tube life (over a year in most cases).

An overnight system of this kind has added point in a hostel or an old-people's home. The main general priority here is that of avoiding an institutional impression, but it is only realistic to accept that wear and tear are likely to be more severe, and to plan accordingly. In a student's room, for instance, it may make for initial economy to provide one portable fitting that can be used at the desk or standing on a shelf at the bed, but the need for repeated physical movement, unplugging and plugging-in again, makes it likely that fairly rapid replacement will be necessary. As a principle, mechanical flexibility is likely to lead to maintenance problems before electrical flexibility does.

Hotel-lighting lies in the no man's land between the commercial and the domestic. Many of the types of interior, such as restaurants, have already been discussed. The treatment generally should help to confirm the character of the hotel in the guest's mind—Hilton or Bristol, country inn or motel? The public rooms are very likely to need differing treatment for daylight hours and evening (anyone designing lighting for a hotel dining-room should have 'remember breakfast' pinned on his board), and it is worth considering a lower level again for the midnight-till-dawn period, particularly in circulation space near bedrooms, where, if nothing else, it will tend to make people move around more quietly. We may take a typical corridor on a private-room floor as an example. It is likely to be without daylight, or have only widely separated windows at each end. One of the functions of the lighting is to break up the linear space into areas of interest. If the doors are grouped in fours, two on each side, as they often are, this defines a specific area, and there may be a small setting back of the wall. The corridor thus becomes a series of entrance zones and links between them. Local tungsten-quality lighting for these zones may be integrated with the fabric (utilising the break in the wall line, perhaps, or changes in ceiling level) or a surface-mounted fitting can help to identify the area. Then, midway between them, a flush recessed fitting can take cool fluorescent tubes. In addition, both lamp housings have a low-intensity source for overnight illumination (this might conveniently also be maintained emergency lighting). The automatic switching pattern would be

	Door zones	Linking corridor
Dawn–dusk	warm/full	cool/full
Dusk–midnight	warm/full	low intensity
Midnight–dawn	low intensity	low intensity

In an all-tungsten scheme, automatic dimming is an alternative possibility.

Within the private-rooms, the feeling is domestic, or idealised domestic, the standardised furniture arrangement simplifying the planning. Any fluorescent lamps should be De Luxe Warm White or Softone27. Switching is important, and refinements such as dimming, or a relay system giving control of both the local light and the general illumination from each point, will impress the visitor. Although it is nearly ten years since it appeared, the special issue of the

International Lighting Review devoted to hotels (Reference 2.76) remains a worthwhile reference; new examples are constantly being reported in architectural and interior-design journals. Hotel-lighting responds readily—perhaps too readily—to the movement of fashion, and can act as a barometer for future trends in other parts of the commercial and domestic fields. Lighting naturally reflects the standards and aspirations of the times, and is part of social history (References 2.77 and 2.78). The proper study of the lighting designer is man.

Postscript

No open-ended and continuously developing subject such as lighting design will admit of comprehensive treatment. Having completed the book, I feel its title might have been longer: 'Electric lighting: some aspects of the design of schemes for spaces within buildings'. Or, as I tried to say in Chapter 19, not 'what you should do is . . .', but rather 'some of the things you might bear in mind are . . .'. Nevertheless, I cannot read the manuscript at this stage without thinking of things that might have been mentioned, or realising that even a short interval brings changes in standards and terminology.

These notes in conclusion, then, have broadly three purposes. The first is to acknowledge recent changes such as new editions of reference works, or in the language that is preferred. The second is to add to the References some titles which have not spontaneously come up in the discussion but which would represent unacceptable omissions even by the arbitrary standard applied here. The third is to look very briefly at broader issues; for, if lighting design is relevant, as we have seen it is, to strategic decisions about buildings, it must relate with those decisions to the total environmental context, nationally and internationally.

In the language of lighting, the major change is that what we have long called 'illumination'—the concentration of incident luminous flux measured in lumens per square metre, or lux—should henceforth be called 'illuminance', the term 'illumination' being applied to the general process of distributing or applying light. The illumination of a room is thus very nearly the same thing as the lighting of a room, and we might say, correctly, if awkwardly, that 'in the illumination of this area, the average illuminance on the working plane is 600lux'. There is an academic justification for the '-ance' ending for a quantity measured in this way, but the possible verbal confusion between 'illuminance' and 'luminance'—to say nothing of 'luminaire' and 'luminous'—means that we shall need to speak clearly as well as think clearly. People will probably go on talking of illumination levels for some time. A revised British Standard on terminology appeared towards the end of 1972 (Reference 2.79).

Changes in emphasis naturally follow developments in lighting equipment. The lamp manufacturers seem to be putting ever more stress on the virtues of high-pressure discharge sources for interior lighting, particularly mercury-halide lamps and improved-phosphor fluorescent-bulb types. In some interiors, such as a hangar for Jumbo jets, there are clear economic advantages (in a hangar at London Airport, high-pressure sodium and mercury have been used in combination) but the colour quality still has limited acceptability, and apart from incidental probiems, such as flicker or cooling and running up again after an interruption in the supply, there remains an essential difficulty with any very powerful concentrated source, in that it is hard to reconcile freedom from glare with adequare illuminance on vertical planes.

In lighting by tubular fluorescent lamps, the relative use of types with good colour rendering continues to grow. Until recently, those who have argued that, even at a fixed loading, in W/m^2, they would prefer the better quality of de luxe tubes and a lower illuminance, have done so on a personal, subjective basis. A recent study by Bellchambers and Godby (Reference 2.80) investigates related choices systematically, and tend to show general sympathy for this reaction. Observers considered a pair of identical small rooms, one lit by a de-luxe type and the other by high-efficacy tubes. Levels were set for 'equal clarity', and, in another series of tests, for 'equal pleasantness'. The settings thought desirable for the better colour-rendering tubes were consistently lower numerically, so that the output advantage of tube types such as Daylight, White, and Warm White, is—at least in some circumstances—more apparent than real. More work on this, particularly in relation to the external daylit scene, is clearly desirable.

The discussion of lighting fittings in Chapter 12 quoted BS3820 (Reference 2.28) as a main reference. This is now being superseded by BS4533, parts 1 and 2 (Reference 2.81), which is being published in sections to a programme due for completion in mid-1973.

A new edition of the IES Code will have appeared before this book is published replacing that of 1968 (Reference 1.3). The draft shows a comprehensive table of illuminance recommendations; they are based both on previous recommendations and a study of good practice in the intervening years, and the values are to represent a 'good standard' rather than minima as previously. The revised scale (in lux) is: 2, 5, 10, 15, 30, 50, 100, 150, 300, 500, 750, 1000, 1500, 3000. But great emphasis is to be placed on the tabulated value being an initial figure to which modifying factors should be applied. Influences such as age, importance of error, and unusual task characteristics appear in a chart leading to a final recommendation. The system is intended to make people using it think about the meaning of what they are doing.

There is, of course, much in the Code beyond the specification of horizontal-plane illuminance, and other recent developments have implied that this is not such an important criterion as may once have been thought. The specialist CIE committee on visual performance presented a report to the 1971 session in Barcelona, Spain, and this was published in 1972 (Reference 2.82). Its main ideas were discussed in a subsequent article by Mardsen (Reference 2.83).

The bibliography and references in this book have been assembled as topics in the developing discussion suggested them. Readers wishing to pursue particular aspects should refer to more systematic lists such as those in the IES Code (Reference 1.3), in 'Lamps and lighting' (Reference 2.29), in 'The ergonomics of lighting' (Reference 2.15), and in the other sources to which they lead. James Bell's section on lighting in the Architectural Press's annual 'Specification' (Reference 2.84) is a valuable summary for the practising designer. It discusses the subject under four main headings: lighting and the form of buildings, good lighting in buildings, lighting in an architectural-design procedure, and calculation techniques. There is a useful bibliography too; it reminds one of an interesting group of books approaching lighting from the architectural viewpoint that appeared about ten years ago mostly from across the Atlantic (e.g. References 2.85, 2.86 and 2.87). Within the last year, we have had the publication of L. C. Kalff's 'Creative light' (Reference 2.88), a group of reflective essays making pleasant and interesting reading for those familiar with the accepted wisdom in this field. Kalff had long ago been responsible for one of the first works seeking to relate lighting comprehensively to architecture (Reference 2.89). Unfortunately, neither this nor a later 3-volume survey of the whole field of lighting technique by his colleague Joh. Jansen (Reference 2.90) has ever appeared in an English version.

Important recent additions to the general literature of lighting include a report on the use of computers (Reference 2.91), and four papers, among others, given at the 1972 national lighting conference of the British IES (Reference 2.92, 2.93, 2.94 and 2.95). A new edition of the American IES handbook has appeared (Reference 2.96).

Derek Phillips and I gave a paper on visual conditions at a recent symposium on the environment in buildings (Reference 2.97). We tried to formulate some answers over particular topics, but we were forced to conclude in uncertainty. 'The major question mark over the development of buildings generally is the extent to which it will be influenced by conservation ideas. Already the case for long-life, loose-fit, low-energy structures is being argued. It is possible we shall look back on the deep-form, technology-intensive building of the seventies as a short-lived historical sport. On the other hand, it may point the way to the best utilisation of energy that can be achieved.'

This is not the place to rehearse the ecological argument. Anyone unfamiliar with it can pick up paperbacks of 'Only one earth' or 'Blueprint for survival' on any bookstall (Reference 2.98 and 2.99). Just one sentence from the latter's introduction may be quoted: 'The principal defect of the industrial way of life with its ethos of expansion is that it is not sustainable.' This has only to be stated for its truth to seem self evident. Indefinite expansion must be incompatible with finite resources. The only difference of opinion is over how soon the crisis will be with us, and there are those who argue it is here and now. *The Observer* of the 17th December 1972 ran the headline 'Energy crisis frightens America', and William Millinship reported that '. . . in Washington the real

Plate 87 Upward light on to decorated ceiling, but brightness on altar reveals that there is also direct light from concealed sources

Plates 88 & 89 Contrasting lighting treatments in galleries with deep baffled ceilings

Plate 90 Entrance area at Skyline Hotel, London Airport

Plate 91 Bedhead lighting at Cumberland Hotel, London

Plate 92 Fully recessed square downlights used here have multi-baffled apertures in general space but acrylic-block insets that give more light toward horizontal at mirror over basins

Plate 93 Luminous ceiling area in this domestic kitchen has panels of small-cell plastic louver screening tubes giving light of tungsten colour quality. Special square mountings are used to relate circular downlights to timbered ceiling

Plate 94 Playroom with track for adjustable spotlights

concern is not about inconvenience this winter, but the long-term effects of the energy crisis'. The director of the Office of Emergency Preparedness, who advises the US President on supplies of strategic materials, was quoted as saying that energy 'is going to replace the cold war as perhaps the most urgent problem America faces in the years ahead. . . . We are entering a new era, an energy-deficit era .. our society, .. our very way of life, its quality and goals, are dependent on how well we meet this challenge.'

It was against a background of growing recognition of the preciousness of expendable resources that Alex Gordon, as president of the Royal Institute of British Architects, instituted the 'long-life, loose-fit, low-energy' study, intended to explore, on many levels, the implications of approaching building design with these general objectives. The difficulty is that the techniques and data we need for that exploration are, at best, incomplete. Conventional accountancy is concerned with the values of the particular national market place, now. True cost, in a global, social sense, is difficult to comprehend and almost impossible to evaluate, but it is the only criterion with ultimate validity. Chapter 15 of this book, on the relation of electric lighting to building form, began by saying that 'the major significance of the development of the fluorescent lamp is that the cost of working illumination is now of the same order whether we provide it electrically or by daylighting'. But we need in the end to consider all the consequential costs of deep planning and depending on relatively high energy consumption for the building's survival as a habitable unit. We certainly do not know the answers yet, but the questions must be put. The Electricity Council in the UK has recently been promoting the claims of the compact-form, sealed, airconditioned building with permanent high-level lighting as the inevitable result of integrated environmental design, and the ideas demand investigation against a background of long-term thinking about resources. The case is a good one when expressed in current costs, and this is not surprising when one remembers that the adoption of buildings with a high technology content by business organisations follows commercial pressures. Barry Commoner summed up a truth that applies to buildings as well as other manmade objects when, at the 1972 RIBA conference (Reference 2.100), he said: 'Changes in design—whether of industrial or agricultural production, transport, individual buildings, or entire urban areas—are governed, not by environmental compatibility, but by the short-term gains which they promise. . . . Environmental deterioration is largely the result of concurrent changes in the nature of productive technologies. . . . The motivation of counter ecological trends in modern technology is economic.' Economic, that is, in terms of profit maximisation; whether the same conclusions are reached if the criterion is the total economy of spaceship *Earth* is another matter.

Does all this mean that we should dust our daylight protractors and begin again to consider the subtleties of splayed reveals in thick, heavy-structure walls? It is too early to say, though, if the 'long-life, loose-fit, low-energy' slogan is a wise one, it does seem to accord with buildings that have, as Pat O'Sullivan has put it, a high natural science content but a low technology content.

More studies are needed of ways of meeting the same basic brief in buildings of different strategy. Projecting costs forward in time for the various stages of a long-life building is hazardous, but a central thread in any analysis of this kind. Apart from the problem of making any estimate of the amount of oil we may have left by, say, 1990, we have also to guess how soon governments will begin to respond to ecological reality. For how long, for instance, will the ludicrous situation persist in the UK, where taxation encourages buildings with a low initial cost and a high running cost? Whether it is conscious policy or not, the kind of construction being fostered could reasonably be described as short-life, tight-fit, high-energy.

To return, finally, to the prospects in lighting design, there is at least a possibility that illuminance levels will fall, particularly those at the top of the range; there is already evidence of a move this way in the USA. It is possible too that the tendency for daytime electric lighting to increase may ultimately be reversed. Streetlighting, someone once said, is like Chinese cookery—the art of making a little going a long way. Perhaps before the century ends, this will be true of interior lighting as well, and the successful lighting designer will be using less light to greater effect. This in no way diminishes the relevance of the ideas that have been discussed in this book—rather the reverse. However sophisticated analysis becomes, it has its limitations. Beyond them the human being, 'the best integrating device yet developed', takes over, and fundamental design decisions come down to value judgments. That, in the end, is what lighting design is about: the values in our visual environment.

Bibliography

1 Essential references

1.1 HOPKINSON, R. G.: 'Architectural physics: Lighting' (HMSO, 1963)
1.2 'Interior lighting design,' Lighting Industry Federation, 1969 (in co-operation with the Electricity Council and the Electrical Contractors' Association)
1.3 IES Code: 'Recommendations for lighting building interiors.' Illuminating Engineering Society, London, 1968
1.4 BS (in preparation): 'Code of basic data for the design of buildings (CP3)—Chap. 1: Lighting. Pt. 2—Artificial light'

2 Further reading

2.1 'Lighting in corrosive, flammable and explosive situations'. Illuminating Engineering Society, London, Technical Report 1, 1965
2.2 'The calculation of utilization factors: the BZ method.' Illuminating Engineering Society, London, Technical Report 2, 1971
2.3 'The lighting of building sites and works of engineering construction.' Illuminating Engineering Society, London, Technical Report 3, 1966
2.4 'Daytime lighting in buildings.' Illiminating Engineering Society, London, Technical Report 4, 1972
2.5 'Lecture theatres and their lighting.' Illuminating Engineering Society, London, Technical Report 5, 1963
2.6 'The floodlighting of buildings.' Illuminating Engineering Society, London, Technical Report 6, 1964
2.7 'Lighting for sport.' Illuminating Engineering Society, London, Technical Report 7, 1965
2.8 'Lighting of libraries.' Illuminating Engineering Society, London, Technical Report 8, 1966
2.9 'Depreciation and maintenance of interior lighting.' Illuminating Engineering Society, London, Technical Report 9, 1967
2.10 'Evaluation of discomfort glare: the IES glare index system for artificial lighting installations.' Illuminating Engineering Society, London, Technical Report 10, 1967
2.11 'The calculation of direct illumination from linear sources.' Illuminating Engineering Society, London, Technical Report 11, 1968
2.12 'Hospital lighting.' Illuminating Engineering Society, London, Technical Report 12, 1968
2.13 'Industrial area floodlighting.' Illuminating Engineering Society, London, Technical Report 13, 1969
2.14 'Lighting of art galleries and museums.' Illuminating Engineering Society, London, Technical Report 14, 1970
2.15 HOPKINSON, R. G., and COLLINS, J. B.: 'The ergonomics of lighting.' (Macdonald, 1970)
2.16 COOMBER, D. C., and JAY, P. A. (1967): 'A simplified method of calculation for luminance

ratio and designed appearance lighting installations,' IES Monograph 10, Illuminating Engineering Society, London
2.17 LYNES, J. A.: 'Lightness, colour and constancy in lighting design,' *Light. Res. & Technol.*, 1971, **3**, (1), pp. 24–42
2.18 JAY, P. A.: 'Lighting and visual perception,' *ibid.*, 1971, **3**, (2), pp. 133–146
2.19 GREGORY, R. L.: 'Seeing in the light of experience,' *ibid.*, 1971, **3**, (4), pp. 247–250
2.20 CUTTLE, C.: 'Lighting patterns and the flow of light,' *ibid.*, 1971, **3**, (3), pp. 171–189
2.21 LYNES, J. A., BURT, W., JACKSON, G. K., and CUTTLE, C.: 'The flow of light into buildings,' *Trans. Illum. Eng. Soc.*, 1966, *31*, (3), pp. 65–91
2.22 ALDWORTH, R. C., and BRIDGERS, D. J.: 'Design for variety in lighting,' *Light Res. & Technol.*, 1971, **3**, (1), pp. 8–23
2.23 'Electric lamps A: Fundamentals of light and its production.' Lighting Industry Federation, 1967
2.24 'Electric lamps B: Fluorescent lamps.' Lighting Industry Federation, 1967
2.25 'Electric lamps C: Incandescent lamps.' Lighting Industry Federation, 1971
2.26 'Electric lamps D: Discharge lamps.' Lighting Industry Federation, 1968
2.27 BEAN, A. R., and SIMONS, R. H.: 'Light fittings: Performance and design' (Pergamon Press, 1968)
2.28 BS3820: 1964. 'Electric lighting fittings'
2.29 HENDERSON, S. T., and MARSDEN, A. M. (Eds.): 'Lamps and lighting' (Arnold, 1972)
2.30 HOPKINSON, R. G., PETHERBRIDGE, P., LONGMORE, J.: 'Daylighting' (Heinemann, 1966)
2.31 LYNES, J. A.: 'Principles of natural lighting' (Elsevier, 1968)
2.32 'Estimating daylight in buildings. Pts. 1 & 2.' Building Research Station, Digests 41 & 42 (HMSO, 1964)
2.33 'Planning for daylight and sunlight.' Planning Bulletin 5 (HMSO, 1964)
2.34 MANNING, P.: 'The design of roofs for single-storey general-purpose factories.' Department of Building Science, University of Liverpool, 1962
2.35 'A report on the problems of windowless environments.' GLC Research Paper 1, Greater London Council, County Hall, London. 1st edn., 1966; revised edn., 1968
2.36 MCNEILL, G. V.: 'Lighting: the metric 'seventies,' *Light. Res. & Technol.*, 1969, **1**, (3), pp. 148–160
2.37 STECK, B.: 'European practice in the integration of lighting, air conditioning and acoustics in offices,' *ibid.*, 1969, **1**, (1), pp. 8–23
2.38 LANGDON, F. J.: 'Modern offices—a user survey,' National Building Studies Research Paper 41 (HMSO, 1966)
2.39 'Integrated design: a case history.' Electricity Council, 1969
2.40 'Head office for Manweb Electricity.' Merseyside & North Wales Electricity Board, 1970
2.41 'Avonbank, a planned environment.' South-Western Electricity Board, 1971
2.42 BODMANN, H. W.: 'Light and the total energy input to a building', *Light & Lighting*, 1970, **63**, (9), pp. 240–249
2.43 PAGE, J. K.: 'UK practice on integrated environmental design,' *Light. Res. & Technol.*, 1970, **2**, (3), pp. 135–149
2.44 DORSEY, R. T.: 'A unified system for the aesthetic and engineering approaches to lighting,' *Int. Light. Rev.*, 1971, **22**, (3), pp. 71–99
2.45 JAY, P. A.: 'Inter-relationship of the design criteria for lighting installations,' *Trans. Illum. Eng. Soc.*, 1968, **33**, (2), pp. 47–71
2.46 BOUD, J.: 'Lighting for shops, stores, and showrooms.' Lighting Industry Federation, 1969 (in co-operation with the Electricity Council and the Electrical Contractors' Association)
2.47 BOUD, J.: 'Shop, stage, studio,' *Light & Lighting*, 1966, **59**, (11), pp. 308–317
2.48 BOUD, J.: 'Lighting for effect,' *ibid.*, 1971, **64**, (8), pp. 230–238
2.49 A special issue on shop and display lighting, *Int. Light. Rev.*, 1969, **20**, (2)
2.50 LEYRIE, R.: 'Railway station, Grenoble,' *ibid.*, 1968, **19**, (4), p. 149
2.51 DUNN, R. F. H.: 'Lighting in radar viewing rooms', *Light & Lighting*, 1972, **65**, (1), pp. 6–9
2.52 'Electrics 1972–73 handbook of electrical services in buildings.' Electricity Council, 1972
2.53 'Spectral requirements of light sources for clinical purposes.' Medical Research Council Memorandum 43 (HMSO, 1965)
2.54 COCKRAM, A. H., and COLLINS, J. B.: 'Permanent supplementary artificial lighting of deep hospital wards,' *Light. Res. & Technol.*, 1970, **2**, (3), pp. 174–185

2.55 NE'EMAN, E., ISAACS, R. L., and COLLINS, J. B.: 'The lighting of compact plan hospitals,' *Trans. Illum. Eng. Soc.*, 1966, **31**, (2), pp. 37–58
2.56 BENTHAM, F.: 'The art of stage lighting' (Pitman, London, 1969)
2.57 PILBROW, R.: 'Stage lighting' (Studio Vista, London, 1970)
2.58 CORBETT, A.: 'The Crucible Theatre, Sheffield,' *Light & Lighting*, 1972, **65**, (6), pp. 190–195
2.59 BOUD, J.: 'Mercury Theatre, Colchester,' *ibid.*, 1972, **65**, (11), pp. 378–381
2.60 'Recommendations for the provision of emergency lighting in premises.' British Electrical & Allied Manufacturers' Association, 1972
2.61 WALDRAM, J. M.: 'The lighting of Gloucester Cathedral by the "designed appearance" method,' *Trans. Illum. Eng. Soc.*, 1959, **24**, (2), pp. 85–105
2.62 'Lighting and wiring of churches.' Council for Care of Churches, 1961
2.63 'Lighting in schools.' Department of Education & Science Building Bulletin 33 (HMSO, 1967)
2.64 'Integrated design at Eastergate,' *Light. Equip. News*, Feb. 1971, **5**, (2), p. 24
2.65 D. M.: Book review of 'Lighting in schools,' *Light & Lighting*, 1967, **60**, (11), p. 364
2.66 WOOD-ROBINSON, M.: 'School lighting,' *ibid.*, 1966, **59**, (3), pp. 66–72
2.67 MANNING, P. (Ed.): 'The primary school: an environment for education.' Pilkington Research Unit, Department of Building Science, University of Liverpool, 1967
2.68 MANNING, P. (Ed.): 'Office design: a study of environment.' Pilkington Research Unit, Department of Building Science, University of Liverpool, 1965
2.69 NUNN, J. P., and JONES, R. F.: 'BBC Midland Region Broadcasting Centre,' *Light & Lighting*, 1972, **65**, (1), pp. 2–5
2.70 A special issue on museum and art gallery lighting, *Int. Light. Rev.*, 1964, **15**, (5–6)
2.71 RIBEIRO, J. S., ALLEN, W. A., and DE AMORIM, M.: 'Lighting of the Calouste Gulbenkian Museum,' *Light. Res. & Technol.*, 1971, **3**, (2), pp. 79–98
2.72 PHILLIPS, D.: 'Lighting' (Macdonald, 1966)
2.73 BOUD, J.: 'Lighting for life' (George Godwin, 1962)
2.74 DURRANT, D. W.: 'The millstone round our neck,' *Light & Lighting*, 1972, **65**, (9), pp. 293–297
2.75 'Generic plans: two and three storey houses.' National Building Agency, London, 1965
2.76 A special issue on hotel lighting, *Int. Light. Rev.*, 1963, **14**, (6)
2.77 O'DEA, W. T.: 'Social history of lighting' (Routledge & Kegan Paul, 1958)
2.78 BOUD, J.: 'Up-staging the Joneses,' *Light & Lighting*, 1972, **65**, (9), pp. 298–300
2.79 BS4727: 1972. 'Lighting technology terminology'
2.80 BELLCHAMBERS, H. E., and GODBY, A. C.: 'Illumination, colour rendering and visual clarity,' *Light. Res. & Technol.*, 1972, **4**, (2), pp. 104–106
2.81 BS4533. 'Electric luminaries (lighting fittings). Pt. 1:1971—General requirements and tests; Pt. 2—1972—Detail requirements'
2.82 'A unified framework of methods for evaluating visual performance aspects of lighting.' Commission Internationale de l'Eclairage, publication 19, Paris, 1972
2.83 MARSDEN, A. M.: 'Visual performance—CIE style,' *Light & Lighting*, 1972, **65**, (4), pp. 132–135
2.84 'Specification 1972—Vols. 1 & 2' (Architectural Press, 1972)
2.85 KOHLER, W., and LUCKHART, W.: 'Lighting in architecture' (Reinhold, 1959)
2.86 FLYNN, J., and MILLS, S.: 'Architectural lighting graphics' (Reinhold, 1962)
2.87 PHILLIPS, D.: 'Lighting in architectural design' (McGraw-Hill, 1964)
2.88 KALFF, L. C.: 'Creative light.' Philips Technical Library (Macmillan Press, 1972)
2.89 KALFF, L. C.: 'Kunstlicht en Architectuur' (Meulenoff, Amsterdam, 1941)
2.90 JANSEN, J.: 'Beleuchtungstechnik.' Philips Technical Library, Eindhoven, 1954
2.91 'Use of computers in lighting,' Proceedings of the Illuminating Engineering Society symposium, London, 1972
2.92 WALDRAM, J. M.: 'A review of lighting progress,' *Light. Res. & Technol.*, 1972, **4**, (3), pp. 129–138
2.93 MARSDEN, A. M.: 'What do we want from our lighting?' *ibid.*, 1972, **4**, (3), pp. 139–150
2.94 MORRIS, A. S.: 'Value for money—building economics,' *ibid.*, 1972, **4**, (4), pp. 215–222
2.95 CUNDALL, G. P.: 'Value for money—interior lighting,' *ibid.*, 1972, **4**, (4), pp. 223–235
2.96 KAUFMAN, J. E. (Ed.): 'IES lighting handbook.' Illuminating Engineering Society, New York, 1972, 5th edn.
2.97 PHILLIPS, D. R. H., and BOUD, J. K.: 'Visual conditions for working environments within

buildings.' Proceedings of the international conference on the environment in buildings, Loughborough University, Loughborough, Leics., England, 1972 (to be published)
2.98 WARD, B., and DUBOS, R.: 'Only one earth,' (Penguin, 1972)
2.99 GOLDSMITH, E., ALLEN, R., ALLABY, M., DAVOLL, J., and LAWRENCE, S.: 'A blueprint for survival' (Penguin, 1972)
2.100 COMMONER, B.: 'Alternative approaches to the environmental crisis,' *RIBA J.*, 1972, **79**, (10), pp. 423–429

Glossary

The informal notes below are intended as a modest supplement to the text. It is hoped that the meaning of most terms in this book is either already understood or emerges from their use; the index acts as a guide to related references. For further explanation the reader may turn to the glossary of terms used in interior lighting given in the 1973 editions of both IES Code and 'Interior lighting design'. These in their turn are based on BS4727 : Pt. 4 : 1971/2 'Glossary of terms peculiar to lighting and colour', and on the 3rd edition of the International Lighting Vocabulary issued jointly by the Commission Internationale de l'Éclairage and the International Electrotechnical Commission.

accommodation — The eye's ability to focus on objects at different distances.

adaptation — The eye's ability to adjust its sensitivity for scenes of different brightness.

apostilb (asb) — A metric unit of luminance useful in lighting design although not as an approved SI unit; a surface with a luminance of 1 asb emits 1 lumen from 1 square metre.

batten fitting — A basic fitting for fluorescent lamps consisting of a channel housing the control gear and supporting the lampholders, leaving the tube exposed. In texts of Continental origin, sometimes known as a 'mounting rail'. In stage lighting, a batten is a linear assembly of small floodlights.

black body — A 'full radiator' is a theoretical surface which emits the maximum possible radiation at any given temperature. Such a surface would also be a perfect absorber of incident light, and so is known, colloquially, as a black body.

brightness — The everyday use of this term extends into the language of lighting. It is often desirable, however, to distinguish between apparent or subjective brightness (which is influenced by adaptation and visual experience) and objective luminance (which see).

BZ number One classification of lighting fittings is in terms of the concentration of their output below the horizontal There are ten classes from BZ1 (the most concentrated) to BZ10 (the most dispersive).

candela (cd) The SI unit of luminous intensity, a more precisely defined equivalent of 'candle power'. In lighting design we are more often concerned with luminous flux measured in terms of lumens. A uniform point source of intensity 1 cd emits 4π lumens into the total 3-dimensional space around it, or 1 lumen into unit solid angle.

colour appearance The apparent colour of a source of light or of a white surface illuminated by it.

colour rendering Term applied to a light source to describe how it affects the appearance of the coloured surfaces it illuminates (the US equivalent is 'colour rendition').

colour temperature If a colour matches the appearance of a black body at a particular temperature, that colour can be so described, as when we identify the colour appearance of a particular photographic lamp by saying that it has a colour temperature of 3200 K.

daylight factor A measure of daylight penetration within buildings. If the daylight factor at a point is 2 per cent the illuminance there (usually on a horizontal plane) is that percentage of the simultaneous exterior illuminance from the unobstructed sky (in a strict definition, the luminance distribution over the sky must be specified—the daylight factor for a 'uniform sky' will differ from that for a 'CIE standard overcast sky').

diffuse reflection A perfectly matt surface exhibits diffuse reflection characteristics.

direct ratio The ratio of the flux directly incident on the working plane to the total emitted below the horizontal by a regular array of the fittings type in question. Direct ratio increases with room size, and the graph of its value against room index is the basis for the BZ classification.

electroluminescence
A phenomenon which can serve as the basis of a light source (cf fluorescence and incandescence). If a phosphor such as zinc sulphide is embedded in a dielectric layer between two conductor plates the application of an alternating electric field results in light emission. Devices of this kind have been developed for luminous indicators and signs, but they have a very limited value for illumination.

fluorescence
The property of absorbing light of one colour and emitting light of another, or in the case of most phosphors used in lighting, absorbing ultraviolet and emitting visible light.

full radiation
See 'black body'.

general lighting service (GLS) lamp
A lamp intended for widespread application in lighting schemes—usually a tungsten lamp with a conventional pear-shaped bulb.

glare
The visual discomfort or disability which results from excessive brightness contrasts within the field of view.

glare index
A numerical expression of the potential glare discomfort from a lighting installation.

illuminance, illumination
The preferred term for the concentration or density of luminous flux incident at a surface (lumens per square metre) is now 'illuminance'. The earlier term 'illumination' is still widely used in this sense, but the official view is that it should now mean the process of lighting a surface or a space.

illumination vector
The magnitude and direction of the flow of light at a point in space. If one imagines a plane including the point being turned until the greatest possible difference between the illuminances on each side appears, then that difference is the magnitude of the vector and a line through the point perpendicular to the plane is its direction.

incandescence
As the temperature of a body is raised it emits radiation of progressively shorter wavelengths. When this spreads into the visible spectrum the body is said to be incandescent. In a tungsten lamp the filament is raised to incandescence by the heating effect of the electric current passing through it.

infrared
Electromagnetic radiation beyond the red end of the visible spectrum.

light output ratio — The ratio of the light output of a lighting fitting to that of the lamp it houses.

lumen (lm) — SI unit of luminous flux. The flux emitted within unit solid angle by a uniform point source of 1 candela.

luminaire — Long-standing American term for lighting fitting now officially preferred internationally.

luminance — Term expressing the luminous flux emitted by unit area of luminous or reflecting surface—hence 'objective brightness'. The approved SI unit of luminance is the candela per square matre (cd/m^2), but for practical design purposes the apostilb (which see) is often more convenient.

luminous efficacy — The number of lumens emitted by a lamp for each watt consumed—hence the unit lm/W. The older term 'luminous efficiency' may still be encountered.

luminous flux — The flow of light through space, measured in lumens. It is comparable with the flow of electricity (amperes) or water (litres per second). Quantity of light is cumulative (cf ampere-hours) and is rarely relevant in lighting design, except where some cumulative effect, such as fading of dyes, is involved.

lux (lx) — SI unit of illuminance. When 100lm are incident on $5 m^2$, the average illuminance is 20 lux.

maintenance factor — An expression of the effect of dirt in reducing the light delivered by an installation; the ratio of the average illuminance in practice to what would be received from the same installation if perfectly clean.

mounting height (H_m) — The vertical separation of lighting fitting and working plane.

polar curve — In general, a graph plotted using polar coordinates (r, θ, i.e. distance of point from origin and angle between line joining origin to point and some reference direction). In lighting a polar curve is such a graph showing how the luminous intensity of a source varies with direction.

profile spot — Spotlight with gate for masking shutters or cut-out plate permitting control of size and shape of projected beam.

reflectance

Ratio of flux reflected from a surface to that incident on it. The older term 'reflection factor' may still be met.

room index (k_r)

Single number expressing room proportions, and defined by

$$k_r = \frac{\text{length} \times \text{width}}{H_m\,(\text{length} + \text{width})}$$

where H_m is mounting height of the lighting fittings. For a scheme with indirect or semi-direct fittings the corresponding 'room ratio', with ceiling height above working plane instead of mounting height, may be more useful.

scalar illuminance

Alternative name for mean spherical illuminance, which for a point in space is the average illuminance over the surface of a small sphere surrounding the point.

SI units

The Systeme Internationale d'Unites, or SI, is a rationalised set of metric units being adopted internationally.

spacing/mounting height ratio

Result of dividing the distance from one fitting to the next by their mounting height. For a particular light distribution, the uniformity of illuminance over a working plane is a function of the S/H_m ratio.

specular reflection

Reflection without diffusion, as in a mirror.

stroboscopic effect

If the frequency of flicker of a lamp corresponds with that with which, for instance, one spoke of a rotating wheel takes the place of another, the wheel may appear to be at rest, or rotating at a different speed, or in reverse. The inaccurate or misleading perception of movement due to fluctuating intensities of light sources is known as the stroboscopic effect.

ultraviolet

Electromagnetic radiation beyond the violet end of the visible spectrum.

utilisation factor

The ratio of the total flux arriving at the working plane to the installed flux, i.e. the total lumen output of the lamps in the scheme. The utilisation factor is the central idea in the lumen method of lighting design.

visual acuity

The ability of the eye to discern detail. Visual acuity may be quantified as the reciprocal of the angle subtended at the eye by the smallest object which can be distinguished.

Index

11, 13, 44.

P. 49